W0111629

Pile Design and Construction Practice

Pile Design and Construction Practice

Editor

Raj Patil

Pile Design and Construction Practice

Edited by **Raj Patil**

Printed in 2017

ISBN: 978-1-68117-005-3

Library of Congress Control Number: 2015931815

© 2016 by

SCITUS Academics LLC,
616, Corporate Way, Suite 2, 4766,
Valley Cottage, NY 10989

www.scitusacademics.com

This book contains information obtained from highly regarded resources. Copyright for individual articles remains with the authors as indicated. All chapters are distributed under the terms of the Creative Commons Attribution License, which permits unrestricted use, distribution, and reproduction in any medium, provided the original author and source are credited.

Notice

Reasonable efforts have been made to publish reliable data and views articulated in the chapters are those of the individual contributors, and not necessarily those of the editors or publishers. Editors or publishers are not responsible for the accuracy of the information in the published chapters or consequences of their use. The publisher believes no responsibility for any damage or grievance to the persons or property arising out of the use of any materials, instructions, methods or thoughts in the book. The editors and the publisher have attempted to trace the copyright holders of all material reproduced in this publication and apologize to copyright holders if permission has not been obtained. If any copyright holder has not been acknowledged, please write to us so we may rectify.

Contents

Preface

A guide book for engineering geologists and geotechnical engineers, Pile Design and Construction Practice not only provides a knowledge base, but also gives accurate schematics and calculations, including the resistance of piles to compressive loads and the effects of compressive loading, a myriad of structural designs for all contingencies, focusing on marine constructions and lateral loading. It tackles problems on durability, machinery foundations, underpinning, frozen ground and mining subsidence areas and is a detailed guide to pile testing and ground investigations.

The book summarizes recent changes, including new codified design procedures addressing design parameters and partial safety factors. It also presents several examples, many based on actual problems.

Editor

Chapter 1

Pile Design Practice in Southern Africa Part I: Resistance Statistics

M Dithinde, and J V Retief

Departmert of Civil Ergireerirg, Stellerbosch University, Private Bag XI, Matielard, 7602, South Africa

Departmert of Civil Ergireerirg, Stellerbosch University, Private Bag X1 Matielard Stellerbosch

ABSTRACT

The paper presents resistance statistics required for reliability assessment and calibration of limit state design procedures for pile design reflecting southern African practice. The first step of such a development is to determine the levels of reliability implicitly provided for in present design procedures based on working stress design. Such an assessment is presented in an accompanying paper (please turn to page 72). The statistics are presented in terms of a model factor M representing the ratio of pile resistance interpreted from pile load tests to its prediction based on the static pile formula. A dataset of 174 cases serves as sample set for the statistical analysis. The statistical characterization comprises outlier's detection and correction of erroneous values, using the corrected data to compute the sample moments (mean, standard deviation, skewness and kurtosis) needed in reliability analysis. The analyses demonstrate that driven piles depict higher variability compared to bored piles, irrespective of materials type. In addition to the above statistics, reliability analysis requires the theoretical probability distribution for the random variable under consideration. Accordingly it is demonstrated that the lognormal distribution is a valid theoretical model for the model factor. Another key basis for reliability theory is the notion of randomness of the basic variables. To verify that the variation in the model factor is not explainable by deterministic variations in the database, an investigation of correlation of the model factor with underlying pile design parameters is carried out. It is shown that such correlation is generally weak.

INTRODUCTION

Geotechnical design is performed under a considerable degree of uncertainty. The two main sources of this uncertainty include: (i) Soil parameter uncertainty and (ii) calculation model uncertainty. Soil parameter uncertainty arises from the variability exhibited by properties of geotechnical materials from one location to the other, even within seemingly homogeneous profiles. Geotechnical parameter prediction uncertainties are attributed to inherent spatial variability, measurement noise/random errors, systematic measurements errors, and statistical uncertainties. Conversely, model uncertainty emanates from imperfections of analytical models for predicting engineering behavior. Mathematical modelling of any physical process generally requires simplifications to create a useable model. Inevitably, the resulting models are simplifications of complex real-world phenomena. Consequently there is uncertainty in the model prediction even if the model inputs are known with certainty.

For pile foundations, previous studies (e.g. Ronold & Bjerager1992; Phoon & Kulhawy 2005) have demonstrated that calculation model uncertainty is the predominant component. One of the fundamental objectives of reliability-based design is to quantify and systematically incorporate the uncertainties in the design process. The current state of the art in the quantification of model uncertainty associated with a given pile design model entails determining the ratio of the measured capacity to theoretical capacity. Accordingly, in this paper a series of pile performance predictions by the static formula are compared with measured performances. To capture the distinct soil types for the geologic region of southern Africa, as well as the local pile design and construction experience base, pile load tests and associated geotechnical data from the southern African geologic environment are used.

In reliability analysis and modelling, both materials properties and calculation model uncertainties are incorporated into a performance function representing the limit state design function in terms of basic variables which express design variables (loads, material properties, geometry) as probabilistic variables. The objective of this paper is to present detailed statistical characterisation of model uncertainty for pile foundations. The analysis is an extension of the model uncertainty characterisation reported by Dithinde *et al* (2011). The purpose of the characterisation is to relate southern African pile foundation design practice to reliability-based design as it has been developed and standardised for geotechnical and structural design. The derived statistics constitute the backbone of all subsequent pile foundations limit state design initiatives in southern Africa. Specific usage of the derived statistics include: assessment of reliability indices embodied in the current southern African pile design practice, as presented in the accompanying paper (Retief & Dithinde 2013 - please turn to page 72); derivation of the characteristic model factor for pile foundations design in conjunction with SANS 10160-5 (2011); and reliability calibration of resistance factors. The following topics are presented subsequently:

The geotechnical background to the dataset is briefly reviewed, including the basis and application for classification into homogeneous datasets and the formal definition of the model factor M to represent model uncertainty.

- An assessment and detection of outliers and correction of erroneous samples, considering the sensitivity of reliability analysis to even a limited number of such values in a dataset.
- Using the corrected data and conventional statistical methods to compute the sample moments: mean, standard deviation, skewness and kurtosis for the respective datasets.
- Verification of randomness of the dataset through investigation of any systematic dependence on the relevant design variables.
- Determination of the appropriate probability distribution to represent model uncertainty provides the final step in characterising model factor statistics.

PILE LOAD TEST DATABASE

Although this paper primarily considers the statistical characteristics of southern African pile model uncertainty, as based on the database of model factors reported by Dithinde *et al* (2011), with additional background provided by Dithinde (2007), it is also necessary to appreciate the geotechnical basis and integrity of the dataset. This section presents an extract of the way in which the dataset has been compiled and a formal definition of a model factor *(M)*.

The database of static pile load tests reported by Dithinde *et al* (2011) include information on the associated geotechnical data, such as soil profiles, field and laboratory test results. A comprehensive range of soil conditions, pile geometry and resistance is incorporated in the dataset, to provide extensive representation of southern African pile construction practice. Although the pile load test reports were collected from various piling companies in South Africa, a significant number of pile tests were performed in countries such as Botswana, Lesotho, Mozambique, Zambia, Swaziland and Tanzania. The main pile types in the database include Franki (expanded base) piles, Auger piles, and Continuous Flight Auger (CFA) piles. In addition, there are a few cases of steel piles and slump cast piles. The steel piles are mainly H-piles, with one case where a steel tubular pile was used.

The collected pile load test data was carefully studied in order to evaluate its suitability for inclusion in the current study. For each load test, emphasis was placed on the completeness of the required information, including test pile size (length and diameter), proper record of the load-deflection data, and availability of subsurface exploration data for the site. Only cases where sufficient soil data was available for the prediction of pile resistance were included in the database.

The pile load tests were used to determine the measured pile resistance, while the geotechnical data was used to compute the predicted resistance. The measured resistances from the respective load-settlement curves were interpreted on the basis of Davisson's offset criterion (Davison 1972). However, for working piles, Chin's extrapolation (Chin 1970) was carried out prior to the application of the Davisson's offset criterion. The predicted resistance was based on the classic static formula which is essentially the generic theoretical pile design model based on the principles of soil mechanics. The soil data that was obtained from the survey, and used for the predicted resistance, was mainly in the form of borehole log descriptions and standard penetration (SPT) results. Soil design parameters were selected on the basis of common southern African practice (Dithinde 2007).

Model Factor Statistics

The primary output of the database of pile load tests reported by Dithinde *et al* (2011) consists of the interpreted pile resistance *(Q$_i$)* and the predicted pile resistance (Qp) from which a set of observations of the Model Factor (M) as given by Equation [1] can be obtained:

$$M = \frac{Q_i}{Q_p}$$

(1)

Where:

Q_i = pile capacity interpreted from a load test, to represent the measured capacity;

Q_p = pile capacity generally predicted using limit equilibrium models, and *M* = model factor.

Each case of pile test included in the dataset is consequently treated as a sample of the set of *n* cases under consideration. In Dithinde *et al* (2011) the complete set of 174 cases was further classified in terms of four theoretical principal pile design classes based on both soil type and installation method. These fundamental sets of classes include: (i) driven piles in non-cohesive soil (D-NC) with 29 cases, (ii) bored piles in non-cohesive soil (B-NC) with 33 cases, (iii) driven piles in cohesive soils (D-C) with 59 cases, and (iv) bored piles in cohesive soils (B-C) with 53 cases. In this paper, the principle four data sets are now combined into various practical pile design classes considered in design codes such as SANS 10169-5 (2011) and EN 1997-1 (2004). The additional classification schemes include:

- Classification based on pile installation method irrespective of soil type. This is the classification adopted in EN 1997-1 (2004) and it yields: 87 cases of driven piles (D) and 83 cases of bored piles (B).

- Classification based on soil type. This classification system is supported by the general practice where a higher factor of safety is applied to pile capacity in clay as compared to sand. This combination results in 58 cases in non-cohesive soil (NC) and 112 cases in cohesive soil (C).

- All pile cases as a single data set irrespective of pile installation method and soil type. This is the practical consideration presented in SANS 10160-5 (2011) where a single partial factor is given for all compressive piles. The scheme yields 174 pile cases (ALL).

DETECTION OF DATA OUTLIERS

The presence of outliers may greatly influence any calculated statistics, leading to biased results. For instance, they may increase the variability of a sample and decrease the sensitivity of subsequent statistical tests (McBean & Rovers 1998). Therefore prior to further numerical treatment of samples and application of statistical techniques for assessing the parameters of the population, it is absolutely imperative to identify extreme values and correct erroneous ones.

The statistical detection and treatment of outliers in the principal four sets were reported by Dithinde et al (2011). The methods used include (i) load-settlement curves, (ii) sample z-scores, (iii) box plots, and (iv) scatter plots. The results for cases with outliers are reproduced in Figure 1. Inspection of Figure 1(a) reveals two potential outliers (i.e. cases 27 and 54). The curves for these two cases depict different behaviour from the rest of the curves (case 27 with a soft initial response and case 54 with a large normalised capacity). Visual inspection of Figure 1(b) for outliers shows one data point marked as outlier for B-NC and B-C data sets. The tagged data points correspond to pile cases number 53 and 156. However, it should be noted that the box plot method for identifying outliers has shortcomings, particularly for small sample sizes as is the case here. Accordingly the identified cases will have to be corroborated by other methods. Examination of Figure 1(c) shows two data points with z-score values at a considerable distance from the rest of the data points. These data points belong to B-NC (case 53) and B-C (case 156) with z-scores of 3.13 and 2.95 respectively. Although the z-score for case 156 is less than the criterion limit value of 3, and therefore technically is not an outlier, it is sufficiently close to the limit to require further scrutiny. The results of the scatter plots of pile capacity (Q_i) versus the predicted capacity (Q_p) revealed the same outliers detected by the other methods.

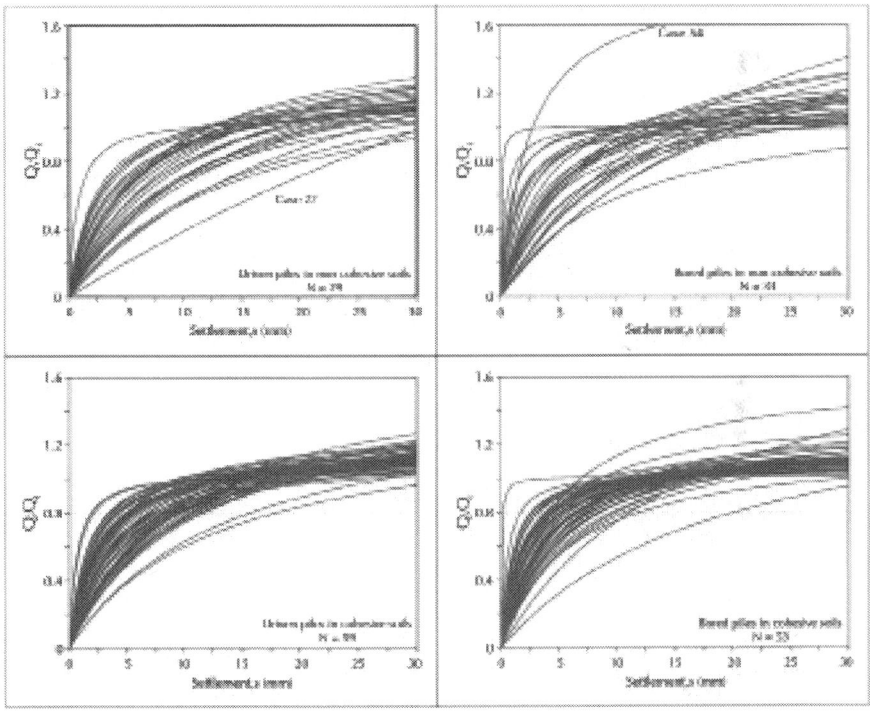

Figure 1: (a) Load-settlement curves method.

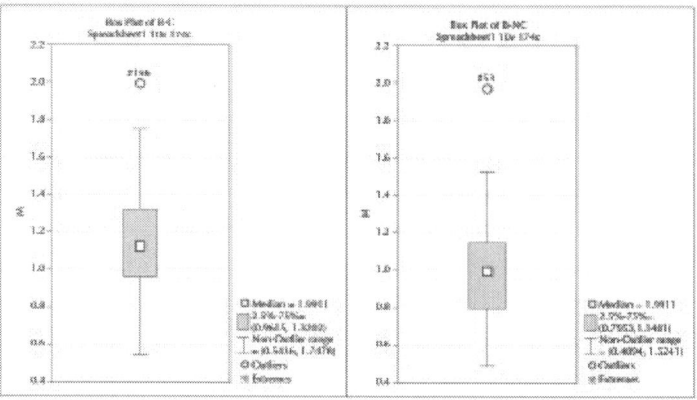

Figure 1: (b) Box plot methods.

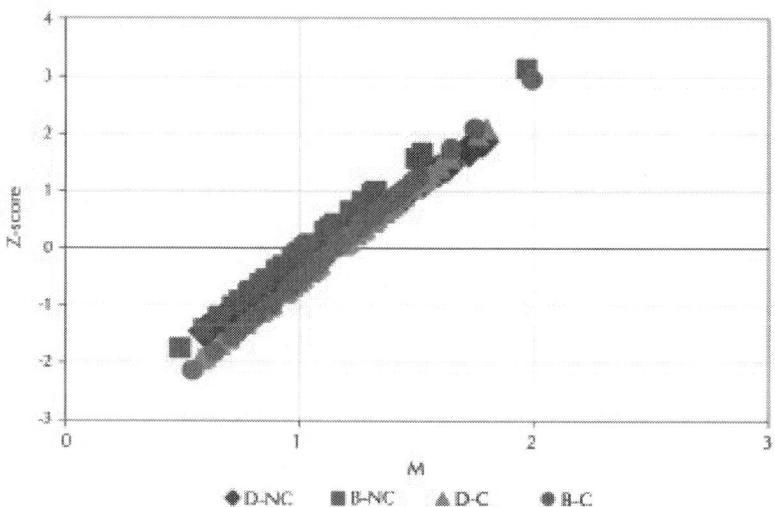

Figure 1: (c) Z-score method.

Aggregate of Outliers

A total of five observations were detected as potential outliers, namely cases 27, 53, 54, 55 and 156. However, it is not proper to automatically delete a data point once it has been identified as an outlier through statistical methods (Robinson *et al* 2005). Since an outlier may still represent a true observation, it should only be rejected on the basis of evidence of improper sampling or error. Accordingly the five data points identified as outliers were carefully examined by double-checking the processes of determination of interpreted capacities and computation of predicted capacities. This entailed going back to the original data (pile testing records and derivation of soil design parameters) and checking for recording and computational errors. Following this procedure the corrections were as follows:

- **Cases 53, 54 and 55:** Examination of records for these cases showed that an uncommon pile installation practice was employed. The steel piles were installed in predrilled holes and then grouted. The strength of the grout surrounding the piles contributed to the high resistance and hence the higher interpreted capacities. Since the installation procedure for these piles deviates from the normal practice, they represent a different population. These were the only piles in the database constructed in this rather unusual method. These data points were therefore regarded as genuine outliers and were deleted from the data set.

- **Case 27:** There was no obvious physical explanation for the behaviour of pile case 27. The depicted behaviour is attributed to extreme values of the hyperbolic parameters representing the non-linear behaviour of the test results. Since piles in terms of pile type, size and soils conditions (i.e. cases 28 and 29) did not show similar characteristics, it was concluded that an error was made during the execution of the pile test. Accordingly this pile case was regarded as having incomplete information, and was therefore deleted.

- **Case 156:** Again there was no obvious physical explanation for the behavior of this pile case. Furthermore, the location of this data point on the scatter plot of Qi versus Qp fits the general trend for the dataset. Therefore no correction was justified for this pile case.

In summary, four outliers were removed and one retained, bringing the dataset to $n = 170$ cases in total and for the respective subsets ^{N}D-NC $= {}^{28}$; ^{n}ab-NC $= {}^{30}$; ^{N}D-C $= {}^{59}$; ^{n}ab-C $= {}^{53}$.

SCATTER PLOTS OF Q$_I$ VS Q$_P$

Scatter plots of Q_i versus Q_p can serve as a multivariate approach to outlier detection. However they are presented here to provide an indication of whether the variance of the data set is constant or varies with the dependent variable (i.e. homoscedasticity). The ensuing scatter plots are presented in Figure 2. Visual inspection of the scatter plots seems to suggest variation in the degree of scatter increases with values of Qp. In this regard, it is evident that there is reduced scatter at smaller values of Qp. However, due to the small sample size, the case for large values of Q_p is not sufficiently clear to make any firm conclusion. Furthermore, Figure 2 gives the impression that the variance of the points around the fitted line increases linearly, thereby suggesting that the standard deviation increases with the square root of the values of Q_p. This explains why the scatter tends to flatten off for large values of Q_p. The foregoing assumption implies that weighted regression analysis must be used to establish the relationship between Q_i and Q_p. Such regression analysis was applied in this study (Figure 2) with the regression line forced to pass through the origin. In this case, the slope of the regression line is an estimate of the model factor M.

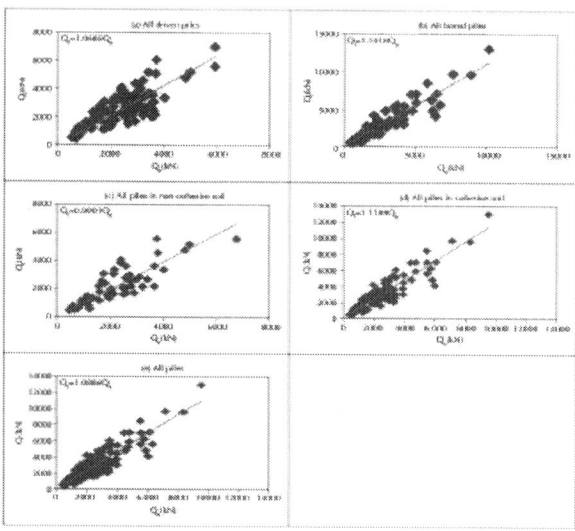

Figure 2: Scatter plots of Q_i Versus Q_p.

SUMMARY STATISTICS

Following the outlier detection and removal process, the descriptive statistics for M consisting of mean (m_M), standard deviation (s_M), skewness and kurtosis are presented in Table 1. The sample descriptive statistics were computed using conventional statistical analysis approach. These are quantities used to describe the salient features of the sample and are required for calculations, statistical testing, and inferring the population parameters.

Table 1: Summary of statics of M

M	n	Mean m_M	Confidence CI -75% m_M -0.75	Std. Dev. s_M	Upper CI SD 75% $s_M+0.75$	COV	Skewness	Kurtosis
D-NC	28	1.11	1.03	0.36	0.40	0.33	0.35	-1.15
B-NC	30	0.98	0.93	0.23	0.26	0.24	0.14	-0.19
D-C	59	1.17	1.12	0.3	0.32	0.26	-0.01	-0.74
B-C	53	1.15	1.10	0.28	0.30	0.25	0.36	0.49
D	87	1.15	1.11	0.32	0.34	0.28	0.1	-0.95
B	83	1.09	1.05	0.28	0.30	0.25	0.41	0.47
NC	58	1.04	1.00	0.30	0.32	0.29	0.55	-0.37
C	112	1.16	1.13	0.29	0.30	0.25	0.15	-0.29
ALL	170	IA.	1.07	0.31	0.32	0.28	0.24	-0.75

The sample mean m_M indicates the average ratio of Q_i to Q_p, with $m_M > 1$ indicating a conservative bias of Q_i exceeding Q_p. This is generally the case, with a positive bias of between 1.04 and 1.17 shown in Table 1, except for the B-NC case where Q_i is on average slightly less than Q_p with $m_M = 0.98$, which is slightly un-conservative. The general conservative bias reflected by m_M is, however, small in comparison to the dispersion of M as reflected by the sample standard deviation S_m for which values range from 0.23 to 0.36; the dispersion is also presented in normalised form as the coefficient of variation $V_M = s_M/m_M$.

The combined effect of values of m_m close to 1 and the relatively large values of S_m or V_m indicate large probabilities of realisations of M in the un-conservative range $M < 1$. The lower tail of the distribution of M derived from the dataset and M statistics is therefore of specific interest for application of the results in reliability assessment.

A comparison of the standard deviations or coefficient of variations for the respective cases indicates small differences. However, there seems to be a distinct trend that is influenced by the pile installation method (i.e. driven or bored). In this regard, driven piles depict higher variability compared to bored piles, irrespective of soil type. This implies that the densification of the soil surrounding the pile emanating from the pile driving process is not well captured in the selection of the soil design parameters. Even the bias for the driven piles dataset is relatively higher, thereby reiterating the notion that current practice is conservative in selecting design parameters for driven piles. Furthermore, the variability in non-cohesive materials is higher than that in cohesive materials. This is attributed to the fact that in cohesive materials the un-drained shear strength derived from the SPT measurement is directly used in the computation of pile capacity, while in non-cohesive materials, the angle of friction obtained from the SPT measurement is not directly used. Instead, the key pile design parameters in the form of bearing capacity factor (N_q), earth pressure coefficient (k_s) and pile-soil interface friction ( ) are obtained from the derived angle of friction on the basis of empirical correlation, thus introducing some additional uncertainties.

Skewness provides an indication of the symmetry of the dataset. The skewness represented in Table 1 is generally positive, indicating a shift towards the upper tail (conservative) of the values for M. There is, however, no consistent trend amongst the values for the respective datasets. The value of 0.24 for the combined dataset (ALL) could therefore be taken as indicative of the general trend. As a guideline it should be noted that the skewness of the symmetrical normal distribution is 0; for a lognormal distribution it is dependent on the distribution parameters, with a value of 0.83 based on the parameter values for the combined dataset.

Values of kurtosis indicate the peakedness of the data, with a positive value indicating a high peak, and a negative value indicating a flat distribution of the data. Negative values generally listed in Table 1 indicate flat distribution of the data, particularly for driven piles. Since these characteristics can only be captured by advanced probability distributions not generally considered in reliability modelling, kurtosis is not further considered.

In order to provide for uncertainties in parameter estimation, Table 1 also presents the confidence limits of the mean and standard deviation at a confidence level of 0.75; this is the confidence level recommended by EN1990:2002 for parameter estimation for reliability models with vague information on prior distributions. The lower confidence limit of the mean (m_m; -0.75) and the upper confidence limit of the standard deviation (s_m-+0 75) is used to present conservative estimates of the range of parameter estimates.

CORRELATION WITH PILE DESIGN PARAMETERS

Although the mean and standard deviation values presented in Table 1 provide a useful data summary, they combine data in ways that mask information on trends in the data. If there is a strong correlation between M and some pile design parameters (pile length, pile diameter and soil properties), then part of its total variability presented in Table 1 is explained by these design parameters. The presence of correlation between M and deterministic variations in the database would indicate that:

- The classical static formula method does not fully take the effects of the parameter into account.
- The assumption that M is a random variable is not valid.

Reliability-based design is based on the assumption of randomness of the basic variables. Since the model factor is among the variables that serve as input into reliability analysis of pile foundations, it is critical to verify that it is indeed a random variable. This was partially verified by investigating the presence or absence of correlation with various pile design parameters. The measure of the degree of association between variables is the correlation coefficient. The basic and most widely used type of correlation coefficient is Pearson r, also known as linear or product-moment correlations. The correlation can be negative or positive. When it is positive, the dependent variable tends to increase as the independent variable increases; when it is negative, the dependent variable tends to decrease as the independent variable increases. The numerical value of r lies between the limits -1 and +1. A high absolute value of r indicates a high degree of association, whereas a small absolute value indicates a small degree of association. When the absolute value is 1, the relationship is said to be perfect and when it is zero, the variables are independent. For the numerical correlation values in-between the limits a critical question is, "When is the numerical value of the correlation coefficient considered significant?" Several authors in various fields have suggested guidelines for the interpretation of the correlation coefficient. For the purposes of this study an interpretation by Franzblau (1958) is adopted as follows:

- Range of r: 0 to ±0.2 - indicates no or negligible correlation
- Range r: ±0.2 to ±0.4 - indicates a low degree of correlation
- Range r: ±0.4 to ±0.6 - indicates a moderate degree of correlation
- Range r: ±0.6 to ±0.8 - indicates a marked degree of correlation
- Range r: ±0.8 to ±1 - indicates a high correlation

The statistical significance of the correlation is determined through hypothesis testing and presented in terms of a p-value. In this regard, the null hypothesis is that there is no correlation between M and the given design parameter (indicative of statistical independence). A small p-value ($p < 0.05$) indicates that the null hypothesis is not valid and should be rejected. Values for the correlation between M and the respective pile design parameters with the associated p-values are listed in Table 2. The results indicate that $R < 0.4$ for all the pile design parameters and therefore the degree of correlation is low. The associated p-values are generally much greater than 0.05, confirming that the correlation between the model factor and the various pile design parameters is statistically insignificant. Therefore, variations in the model factor are at least not explainable by systematic variations in the key pile design parameters, and a random variable model appears reasonable.

Table 2: Correlation with pile design parameters

Design parameter	Case	Spearman rank correlation	
		R	p-value
Pile. length	D	0.11	0.29
	B	0.11	0.31
	NC	-0.25	0.05
	C	0.17	0.07
	ALL	0.02	0.75
Shaft diameter	D	0.01	0.92
	B	0.12	0.26
	NC	0.13	0.39
	C	-0.02	0.82
	ALL	0.05	0.53
Base diameter	D	-0.16	0.15
	B	0.05	0.63
	NC	0.00	1.00
	C		0.51
	ALL	-0.03	0.67
ϕ-base	NC	0.19	0.16
ϕ-shaft	NC	0.19	0.16
C_u-base	C	-0.002	0.98
C_u.shaft	C	-0.21	0.02

For visual appreciation of the correlation results in Table 2, some of scatter plots of M versus pile design parameters are shown in Figure 3.

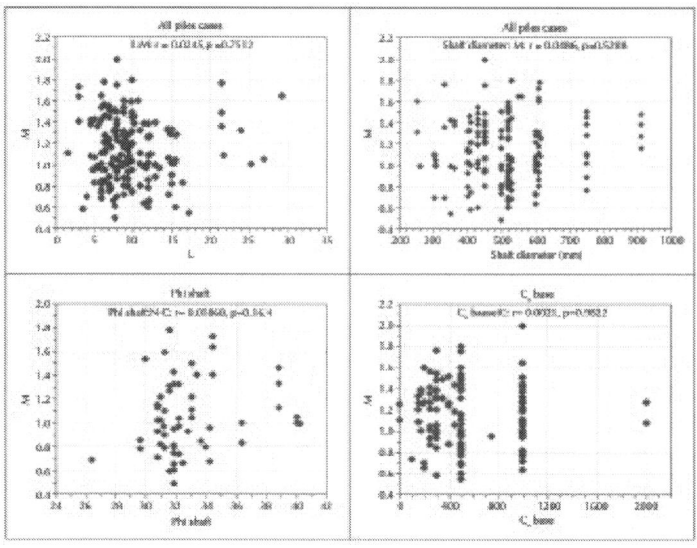

Figure 3: Correlation with some of the pile design parameters.

PROBABILISTIC MODEL FOR THE MODEL FACTOR

The theory of reliability is based on the general principle that the basic variables (actions, material properties and geometric data) are considered as random variables having appropriate types of distribution. One of the key objectives of the statistical data analysis is to determine the most appropriate theoretical distribution function for the basic variable. This is the governing probability distribution for the random process under consideration and therefore extends beyond the available sample (i.e. the distribution of the entire population). Once the probability distribution function is known, inferences based on the known statistical properties of the distribution can be made.

For reliability calibration and associated studies, the most commonly applied distributions to describe actions, materials properties and geometric data are the normal and log-normal distributions (Holický 2009; Allen *et al* 2005). Accordingly, for the current analysis only the normal and lognormal distribution fit to the data are considered. The fit is investigated through (i) a cumulative distribution function (CDF) plotted using a standard normal variate with z as the vertical axis, and (ii) direct distribution fitting to the data.

The cumulative distribution function is the most common tool for statistical characterisation of random variables used in reliability calibration (e.g. Allen *et al* 2005). In the context of the current analysis, the CDF is a function that represents the probability that a value of *M* less than or equal to a specified value will occur. This probability can be transformed to the standard normal variable (or variate), z, and plotted against *M* values (on x-axis) for each data point. This plotting approach is essentially the equivalent of plotting the bias values and their associated probability values on normal probability paper. An important property of a CDF plotted in this manner is that normally distributed data plot as a straight line, while lognormally distributed data on the other hand will plot as a curve. The following steps were used to create the standard normal variate plot of the CDF: ■The capacity model factor values in a given data set were sorted in a descending order, then the probability associated with each value in the cumulative distribution was calculated as $i/(n+1)$.

- For the probability value calculated in Step 1 associated with each ranked capacity model factor value, z was computed in Excel as: $z = NORMSINV(i/(n+1))$ where *i* is the rank of each data point as sorted, and *n* is the total number of points in the data set.
- Once the values of z have been calculated, z versus model factor *(X)* was plotted for each data set.

The ensuing plots are presented in Figure 4(a) from which it can be seen that the CDF for the five data sets plot more as curves than straight lines, thereby implying that the data follow a lognormal distribution. A further characterisation entailing fitting predicted normal and lognormal distributions to the CDF of the data sets is carried out. These theoretical distributions are also shown in Figure 4(a). Both distributions seem to fit the data reasonably well. However, with the exception of the bored piles data set, the lognormal distribution has a better fit to the lower tail of the data, which is important for reliability analysis and design.

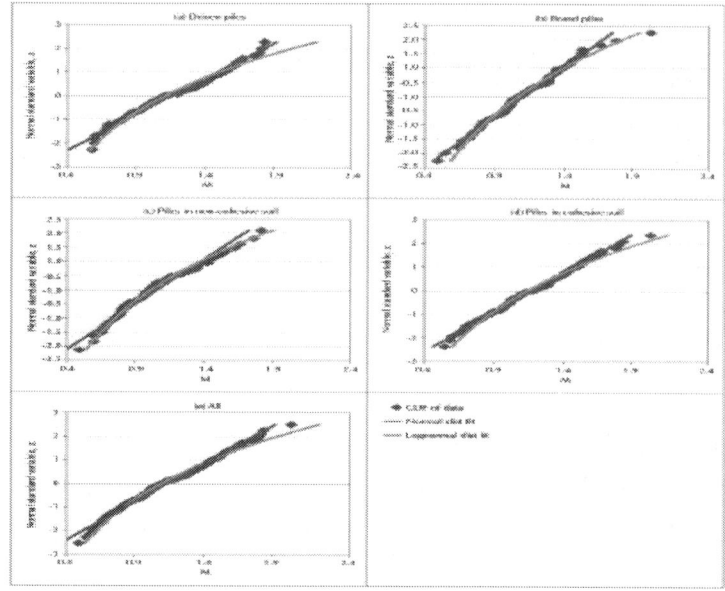

Figure 4: (a) CDF plots with normal and lognormal fit.

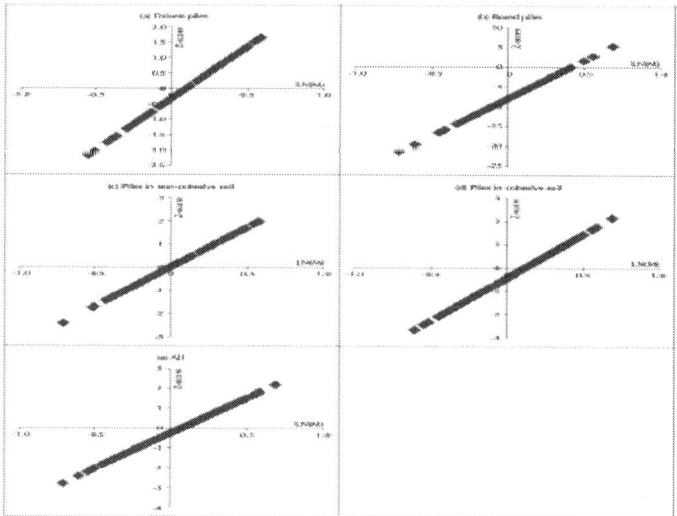

Figure 4: (b) Z-score vs LN (M).

To further confirm that the data best fits a lognormal distribution, z-scores are plotted as a function of Ln *(M)*. The plots would follow a straight line if the data in fact follows the lognormal distribution. The results are presented in Figure 4(b) from which it is apparent that all the data sets plot as a straight line. This therefore confirms the strong case for a lognormal distribution assumption for the data.

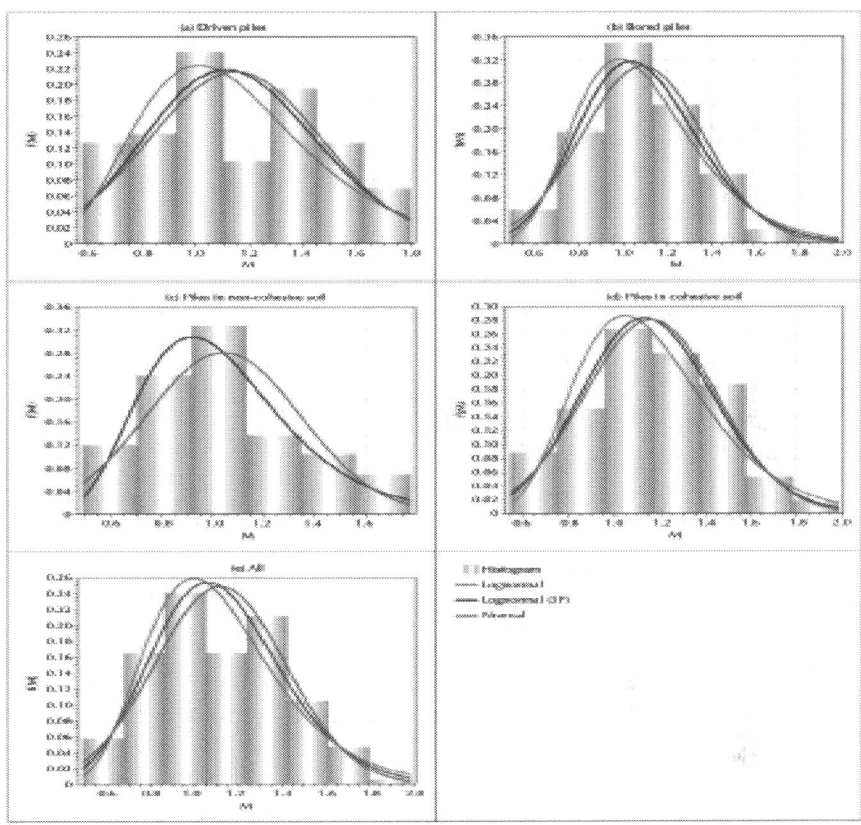

Figure 5: Normal and lognormal distribution fit to the data.

In Figure 5 the histogram of *M* for the respective datasets are compared to normal, lognormal and general lognormal (also three-parameter 3P) probability density function distributions based on the sample moments listed in Table 1 as distribution parameters. The graphic comparison indicates the degree to which the alternative distributions provide a reasonably smoothed representation of the *M* data. At the same time the approximate nature of the M data is indicated by the uneven nature of the histogram. The quantitative assessment of the difference between the empirical data frequencies and the assumed distributions is provided by the chisquare goodness-of-fit test. In this regard, the p-value is a measure of the goodness-of-fit, with larger values indicating a better fit.

In testing the hypothesis that the distribution of the data is similar to the selected probability distribution (normal or lognormal), the hypothesis is rejected if $p < 0.05$. The p-values for chi-square testing are presented in Table 3 from which it is apparent that such values for all the data sets are greater than 0.05 and therefore there is no evidence to reject the null hypothesis of either normal or lognormal distributions. However, on the basis of the magnitude of the p-values, the lognormal distribution seems to show a better fit compared to the other two distributions. The general lognormal distribution provides distributions which are generally intermediate between the normal and lognormal distributions (Figure 5), with similar results for the p-values (Table 3).

On the basis of the results of the two standard distribution fitting approaches studied, it can be concluded that the data fits both the normal and lognormal distributions, although the ordinary lognormal distribution has a slight edge, particularly towards the lower tail (Figure 5). However, theoretically M ranges from zero to infinity, resulting in an asymmetric distribution with a zero lower bound and an infinite upper bound. The lognormal probability density function is often the most suitable theoretical model for such data, as it is a continuous distribution with a zero lower bound and an infinite upper bound. On the basis of this practical consideration, past studies (e.g. Phoon 2005; Briaud & Tucker 1988; Ronold & Bjerager 1992; Titi & Abu-Farsakh 1999; FHWA-H1-98-032 2001; Rahman et al 2002) have recommended the lognormal distribution as the most suitable theoretical model for model uncertainty. Furthermore, in the Probabilistic Model Code by the Joint Committee on Structural Safety (JCSS) (2001), model uncertainty is modelled by the lognormal distribution. Therefore the lognormal distribution is considered a valid probability model for M. Nonetheless, it should be acknowledged that there could be some other distributions that can provide a better fit to the tails. Generally such advanced and complex distributions require a large sample size. For a small sample size, as is the case in this study, such distributions may only lead to a refinement of the results, but not a significant improvement.

CONCLUSIONS

Pile foundation design uncertainties are captured by the M statistics. The M statistics constitute the main input into reliability calibration and associated studies. Since the M statistics are derived from raw data, statistical characterisation of such data is of paramount importance. Accordingly characterisation of the data collected for pile foundation reliability studies have been presented in this paper. The key conclusions reached are as follows:

- Based on the mean values for M, the static formula yields a positive bias of between 1.04 and 1.17, except for the B-NC data set where Qi is on average slightly less than Qp with $mm = 0.98$, which is slightly un-conservative.

- There is a distinct trend that driven piles depict higher variability compared to bored piles, irrespective of materials type. This suggests that the densification induced by pile driving is not fully captured by existing procedures for selecting design parameters.

- The variability in non-cohesive materials is higher than that in cohesive materials. This is attributed to the high degree of empiricism associated with the selection of pile design parameters (N_q, k_s and δ) in non-cohesive soils.

- The values of m_M close to 1 and the relatively large values of S_m or V_m indicate large probabilities of realisations of M in the un-conservative range $M < 1$. Therefore the lower tail of the distribution of M is of specific interest for application of the results in reliability assessment.

- At the customary 5% confidence level, the chi-square goodness-of-fit test results indicate that both the normal and log-normal distributions are valid theoretical distributions for M. However, when taking into account other practical considerations, the lognormal distribution is considered to be the most appropriate distribution for M.

- None of the pile design parameters is significantly correlated with the model factor. From the probabilistic perspective, this implies that the variation in the model factor is not caused by the variations in the key pile design parameters. Therefore it is correct to model the model factor as a random variable.

REFERENCES

1.   Allen, T M, Nowak, A S & Bathurst, R J 2005. Calibration to determine load and resistance factors for geotechnical and structural design. Washington DC: Transport Research Board.

2.   Briaud, J L & Tucker, L M 1988. Measured and predicted response of 98 piles. *Journal of Geotechnical Engineering,* 114(9): 984-1001. Chin, F K 1970. Estimation of the ultimate load of piles not carried to failure. *Proceedings,* 2nd Southern Asian Conference on Soil Engineering, 81-90.

3.   Davison, M T 1972. High-capacity piles. *ProceedingS,* Soil Mechanics Lecture Series on Innovation in Foundation Construction, Chicago, American Society of Civil Engineers, Illinois Section, 22 March, 81-112.

4.   Dithinde, M 2007. Characterisation of model uncertainty for reliability-based design of pile foundations. PhD Thesis, Stellenbosch, South Africa: University of Stellenbosch.

5.   Dithinde, M, Phoon, K K, De Wet, M & Retief J V 2011. Characterisation of model uncertainty in the static pile design formula. *Journal of Geotechnical and Geoenvironmental Engineering,* ASCE, 137(1): 333-342.

6.   European Committee for Standardization 1997. EN-1997:2004. *Eurocode 7: Geotechnical Design. Part 1: General Rules.* Brussels: European Committee for Standardization (CEN).

7.   FHWA (US Federal Highway Administration) 2001. Load and Resistance Factor Design (LRFD) for Highway Bridge Substructures. Publication No FHWA-HI-98-032, Washington DC: FHWA.

8.   Franzblau, A 1958. *A Primer for Statistics for Non-Statisticians.* New York: Harcourt Brace & World.

9.   Holický, M 2009. *Reliability Analysis for Structural Design.* Stellenbosch: SUN MeDIA Press.

10. JCSS (Joint Committee on Structural Safety) PMC 2001. *Probabilistic model code.* JCSS Working Materials [Online] http://www.jcss.ethz.ch (retrieved 1 Feb. 2011).

11. McBean, E A & Rovers, F A 1998. *Statistical Procedures for Analysis of Environmental Monitory Data and Risk Assessment.* Upper Saddle River, New Jersey: Prentice-Hall

12. Phoon, K K & Kulhawy, F H 2005. Characterisation of model uncertainty for laterally loaded rigid drilled shaft.*Geotechnique,* 55(1): 45-54.

13. Phoon, K K 2005. Reliability-based design incorporating model uncertainties. *Proceedings,* 3rd International Conference on Geotechnical Engineering combined with the 9th Yearly Meeting of the Indonesian Society for Geotechnical Engineering, 191-203.

14. Rahman, M S, Gabr, M A, Sarica, R Z & Hossan, M S 2002. Load and resistance factors design for analysis/ design of piles axial capacities. North Carolina State University.

15. Retief, J V & Dithinde, M 2013. Pile design practice in southern Africa. Part 2: Implicit reliability of existing practice. *Journal of the South African Institution of Civil Engineering,* 55(1): 72-79.

16. Robinson, R B, Cox, C D & Odom, K 2005. Identifying outliers in correlated water quality data. *Journal of Environmental Engineering,* 131(4): 651-657.

17. Ronold, K O & Bjerager, P 1992. Model uncertainty representation in geotechnical reliability Analysis. *Journal of Geotechnical Engineering, ASCE,* 118(3): 363-376.

18. SANS 2011. SANS 10160-5:2011: *Basis of Structural Design and Actions for Buildings and Industrial Structures.*Part 5: *Basis for Geotechnical Design and Actions.* Pretoria: South African Bureau of Standards.

19. Titi, H H & Abu-Farsakh, M Y 1999. Evaluation of bearing capacity of piles from cone penetration test data. Project No. 98-3GT, Baton Rouge: Louisiana Transportation Research Centre.

Pile Design Practice in Southern Africa Part 2: Implicit Reliability of Existing Practice

J V Retief; and M Dithinde

Department of Civil Engineering Stellenbosch University Private Bag X1 Matieland 7602 South Africa

Department of Civil Engineering University of Botswana Private Bag UB 0061 Gaberone Botswana

ABSTRACT

Limit state design has become the basis of geotechnical design codes worldwide. With the semi-probabilistic limit state design approach, load and resistance factors of (deterministic) design functions are calibrated on the basis of reliability theory. The calibration is done to obtain procedures that will ensure that a target level of reliability is exceeded under the design conditions within the scope of the design function. This is conventionally expressed in terms of the reliability index ( ), which is related to the probability of failure (P_f). Acceptable existing design practice is an important source of information on appropriate levels of reliability. This paper uses the results from a pile load test database to evaluate the reliability levels implied in the current South African pile design approach. The results of the analysis indicate that the reliability index values for ultimate limit state failure of single piles implicit to present design practice vary with the pile class. However, the influence of the probability model applied is more significant. Based on conventional and standardised procedures for reliability analysis, a representative implicit reliability index value $_{Rep}$ 3.5 is obtained, corresponding to a probability of failure Pf = 2.10^{-4}. The values for various sets of pile conditions range from $_I$= 3.1 $(P_f = 1.10^{-3})$ to $_I$ = 4.3 $(P_f = 1.10^{-5})$. This compares well with target levels of reliability for structural and geotechnical performance of $_T$ = 3.0 as set in SANS 10160-1:2011 Part 1 Basis of structural design. These indicative results provide a useful reference base to establish the reliability of existing and therefore acceptable South African pile design practice. It could also be interpreted as indicative of geotechnical design practice in general. The standard SANS 10160-5:2011 Part 5 Basis for geotechnical design and actions provides the framework for future calibration investigations.

INTRODUCTION

The model factor statistics presented in the accompanying paper (Dithinde & Retief 2013 - please turn to page 60) provide a clear indication of the need for a systematic treatment of the variability and uncertainty of design parameters and procedures in geotechnical design practice. The principles of reliability-based design providing the conceptual basis for such systematic treatment are sufficiently established to be captured in standardised procedures such as the International Standard ISO 2394:1998 (adopted as SANS 2394:2004) and converted into operational basis of design procedures such as Eurocode EN 1990:2002. The standardised procedures are based on limit states design format with a reliability-based framework to ensure appropriate performance levels for the load-bearing capacity and characteristics of the structure or civil engineering works.

Sufficient advances in the theory of reliability have been made to derive guidelines for levels of performance as expressed in terms of reliability representing probability of failure (P_f) for classes of structures and facilities. For various reasons, however, there is insufficient information available to develop reliability-based procedures purely on frequentist or statistical probability models. The most compelling argument for taking information from existing practice into account when reliability-based design procedures are developed comes from the success of present practice and codes which primarily rely on experience-based expertise and judgement.

Capturing the reliability performance from existing practice which is deemed to be acceptable, such as presented in the accompanying paper, is an important source of information for the development of standardised design procedures. One of the possible applications of the information on existing practice is to obtain an indication of acceptable levels of reliability, in comparison to other ways in which target reliability is established. This is the purpose of the present paper. In the process, representative probability models for pile resistance are obtained, that could subsequently be used for reliability calibration of standard design procedures.

A brief overview is firstly provided of the state of limit states design in South Africa -the standardised way in which the principles of reliability is formulated and related to limit states design, including a discussion of target levels of reliability in general and as established for South Africa, serves as basis for comparison of implicit levels of reliability derived for existing practice. Ultimately such implicit levels of reliability are derived, considering alternative probability models from the results of the accompanying paper, and determining representative cases and probability models.

RELIABILITY-BASED GEOTECHNICAL LIMIT STATES DESIGN

The need of converting the now defunct Code of Practice for Pile Foundation Design (SABS 088:1972) to limit states design principles was recognised by the South African piling industry as far back as 1993. A concerted effort was also made by the Geotechnical Division of the South African Institution of Civil Engineering (SAICE) to adopt and apply geotechnical limit state design in South Africa, as summarised by Day & Retief (2009). Recent international and local developments have now added impetus to the introduction of probabilistic-based limit state geotechnical design in South Africa. These include:

i.   Increased interest in harmonisation of technical rules for design of building and civil engineering works internationally as is demonstrated for instance by the activities of the ISSMGE (Orr et al2002) and across disciplines, as demonstrated by the Eurocode set of design standards (CE 2002).

ii.   The international acceptance of semi-probabilistic limit states as the standard basis on which the new generation of geotechnical codes are being developed today, such as Eurocode EN 1997:2004Geotechnical design (EN 1997:2004) and the FHWA Manual for Load and Resistance Factor Design (LRFD) of bridge substructures (FHWA 2001).

iii. The publication of the revised South African Loading Code (SANS 10160:2011 Basis of structural design and actions for buildings and industrial structures) providing the reliability framework in Part 1 Basis of structural design, with the implication that the subsequent materials codes will be based on the same framework. Geotechnical design is included in this framework with the first step taken in Part 5 Basis of geotechnical design and actions as related to buildings and similar industrial structures.

The main advantage of the derivation of geotechnical limit states design procedures from the principles of reliability is that it provides a rational basis for such practice. In addition to enhancing the rationality of design for a specific situation (limit state, failure mode, construction type, etc) it also improves the consistency between the various situations within a single construction (geotechnical, substructure, superstructure, structural materials) or extends to the scope of application of the design procedures.

Common principles of reliability provide the rational basis for unification of geotechnical and structural design. This is an essential requirement for interrelated but specialised design procedures, not only since both elements are shared by individual constructions, but also for the purpose of technical communication between geotechnical and structural design practitioners.

At the highest level a rational basis for the underlying models and procedures is the only way in which international harmonisation of design practice can be maintained. The importance of sharing the wealth of international experience on the basis of harmonisation is usually appreciated, but the ability to provide optimally for local conditions whilst maintaining fundamental alignment with internationally accepted procedures is not always achieved or even attempted.

The theory of reliability, as applied to determine the performance of civil engineering works, is sufficiently mature to formulate standardised procedures for its application in design practice: The International Standard General principles on reliability for structures, ISO 2394:1998, was adopted as a South African National Standard SANS 2394:2004. The Joint Committee on Structural Safety Probabilistic Model Code (JCSS-PMC 2001) provides more detailed pre-normative reliability procedures and models. A notable development is the conversion of general reliability concepts into operational procedures as captured in the Eurocode Standard EN 1990:2002 Basis of structural design which provides a common basis for the set of Eurocodes. SANS 10160-1:2011 and SANS 10160-5:2011 respectively provide harmonisation with the Eurocode for the basis of design and geotechnical design. EN 1990 Annex C Basis of partial factor design and reliability analysis serves as reference for standardised reliability practice, with probabilistic models taken from Annex D Design assisted by testing in this paper.

A critical element of converting reliability analysis into design procedures is the establishment of acceptable levels of reliability. Some guidance on appropriate levels of reliability is given in SANS 2394:2004 and JCSS-PMC: 2001. Application of appropriate levels of reliability in South African structural design is discussed by Retief & Dunaiski (2009). The implicit levels of reliability of existing design practice are recognised in standardised procedures such as SANS 2394:2004, EN 1990:2002 and FHWA HI-98-032 (FHWA 2001) as a basis for selecting target levels of reliability.

MOTIVATION AND PURPOSE OF INVESTIGATION

Given the importance of the reliability performance of existing practice serving as starting point for the calibration of more refined limit states design procedures, the purpose of this paper is to provide such an assessment of present pile construction and design practice in southern Africa. Implicit reliability serves as baseline for acceptable practice. Inconsistency in reliability across the scope of application can be identified, considering possible remedies and adjustments. Systematic calibration of the provisions of SANS 10160-5:2011 is another possible application of the results reported here.

The purpose of this paper is to assess the reliability performance of southern African pile design practice by exploring the application of the database of model uncertainties of pile resistance as reported by Dithinde et al (2011) where particulars of pile load tests and associated geotechnical information, design parameters and descriptive statistics are fully reported. Information from this pile database, together with additional statistical treatment as reported in the accompanying paper, serves as input to the reliability assessment reported here. A comprehensive range of soil conditions, pile geometry and resistance is incorporated in the database, to provide extensive representation of southern African pile construction practice in this assessment.

RELIABILITY CONCEPTS

The concepts of reliability, as developed for geotechnical and structural design, are defined in SANS 2394: 2004. The operational basis for partial factor design and reliability analysis as presented in EN 1990:2002 is followed here since these guidance procedures also apply to SANS 10160-1:2011 for the general basis of design and SANS 10160-5:2011 specifically for the geotechnical basis of design.

The reliability-based performance function for a structure $g(X_1; X_j)$ as a random function of the random variables $(X_1; X_j)$ is expressed as a limit state function indicating the state beyond which the structure no longer satisfies the design performance requirements, as shown in Equation 1. The random variables consist of a specified set of variables representing physical quantities which characterise actions, and material properties (including soil properties and geometrical quantities) conventionally defined as basic variables. The probability of failure of the structure P_f is given by Equation 2. P_f can also conveniently be expressed in terms of the reliability index and the cumulative normal distribution function or the inverse normal distribution function $^{-1}$ as given by Equation 3.

$$g(X_1,...X_j) = 0 \qquad (1)$$

$$P_f = P[g(X_1...X_j) < 0] \qquad (2)$$

$$P_f = \Phi(-\beta) \text{ or } \beta = -\Phi^{-1}(P_f) \qquad (3)$$

Two distinct formats of reliability-based design applies, with Equation 1 representing the probabilistic format. The deterministic partial factors format for standardised limit states design is defined in SANS 101601:2011; the application of the partial factors format to geotechnical limit states design is defined in SANS 10160-5:2011. Reliability calibration is the process of determining appropriate values for the partial factors to achieve a specified level of reliability for a given limit state as derived from Equation 1. Since partial factors design procedures are expressed in deterministic format with various partial factors calibrated on principles or probabilistic reliability, it is classified as a semi-probabilistic procedure or Level 1 reliability-based design (EN 1990:2002).

Although the theory of reliability is firmly rooted in the mathematical theory of probability and related statistics, its success as an operational basis for geotechnical and structural design is directly related to the simplification and approximation applied to the representation of the basic variables (X_i) and solving of the performance function, Equation

1. The ultimate approximation comes from the conversion of Equation 1 into a deterministic design function which employs partial factors that are based on the theory of reliability (see for example Holický et al 2007).

The most important level of approximation is related to the degree to which sources of variability and uncertainty are treated comprehensively. On the one hand it is granted that reliability modelling does not provide for a vital component of failure, which derives mainly from phenomena such as gross human error. Therefore reliability levels are often referred to as notional reliability. On the other hand reliability modelling presents a powerful tool for identification of critical sources of uncertainty, providing the basis for quality management measures in defence against gross error. The most compelling argument for reliability theory to provide for variability and uncertainty is its modelling and predictive capability, equivalent to structural mechanics modelling of load bearing behaviour for structural and geotechnical design.

TARGET LEVEL OF RELIABILITY

Central to the reliability basis of design procedures is the calibration of partial factors, which consists of an inverse reliability analysis process of calculating partial factors to exceed a given or target level of reliability ( ) as an initial step. The establishment of an appropriate level of reliability in accordance with the design case under consideration therefore plays a key role in reliability-based limit states design, or more specifically the calibration of standardised design procedures.

Several approaches for setting the target reliability index are available. A pragmatic approach which is mostly followed is to apply a combination of the various methods. The methods include:

- Risk-based cost-benefit analysis and optimisation
- Failure rates estimated from actual case histories
- Value set by regulatory authorities for a given limit state
- Range of beta values implied in the current design practice.

Risk-based Optimisation of Reliability

The most rational approach for establishing the target level of reliability is through cost-benefit analysis and optimisation. Cost-benefit analysis entails the study of the variation of the initial cost, maintenance costs, and the costs of expected failure. It therefore represents the determination of reliability in the context of risk optimisation. The matter of the necessary inclusion of the loss of human life leads this process to be highly controversial. However, the relatively recent development of the concept of the Life Quality Index (LQI) which relates human life in neutral terms of marginal changes in life expectancy and working life (see for example Rackwitz 2008) should resolve this controversy. Although the LQI concept developed rapidly in recent years, no operational guidelines are available as yet, particularly for South African conditions.

Reliability Levels for Geotechnical Design

The target probability of failure for a given structure can be established on the basis of failure rates estimated from actual case histories. For the case of foundations it is estimated that probability of failure (P_f) ranges from 0.001 to 0.01 - about one-and-a-half orders of magnitude below a "marginally acceptable" level and half an order of magnitude below an "acceptable" level according to the FHWA Manual for LRFD bridge pile design (FHWA 2001). However, many authors (e.g. Phoon 1995; Baecher & Christian 2003; Christian 2004) have cautioned that the probability of failure for constructed facilities is not solely a function of the design process uncertainties, as is the case for the calculated failure probabilities. Therefore, for comparison with calculated failure probabilities, the rate of failure from FHWA (2001) should be adjusted by one order of magnitude downward (Phoon 1995). If the suggested adjustments are effected, the probability of failure for foundations becomes 0.001 to 0.0001 which corresponds to target reliability index values (β_T) of between 3.1 and 3.7.

Reliability Levels for South African Practice

The target levels of reliability for South African constructions within the scope of the revised Loading Code SANS 101601:2011 are discussed by Retief & Dunaiski (2009). Motivation is provided for maintaining the reference level of $_T = 3.0$ to be the same as that applied in SABS 0160:1989 (Milford 1988). The decision was based mainly on the argument that there was no evidence or justification for adjusting the level of reliability for South Africa. The reference reliability agrees with practice in countries such as the USA and Canada. It is consistent with guidance given in SANS 2394:2004 when South African economic conditions are taken into account.

The most serious challenge to maintaining the reference level of reliability for South Africa came from the default value of $_T = 3.8$ applied in Eurocode. It should be noted, however, that this value is not normative in Eurocode, but since safety is treated as a national issue $_T$ can be selected by member countries. The high value of reliability applied in Eurocode was also judged to reflect higher levels of economic development, which implies lower relative cost of construction (or higher affordability) and consequent higher safety levels obtained in risk-based optimisation.

A factor which moderates the difference between South African target reliability and the Eurocode default value is that in calibration the reliability is seen as a constraint, whilst the Eurocode value is often seen as a target to be attained on average (SAKO 1999). The implication is that $_T = 3.0$ as a constraint differs less from $_T = 3.8$ as an average target than it may appear at first glance.

Another moderating factor is that the South African reference value applies to a more restricted reliability class of construction (RC2, typically buildings up to four storeys high) which corresponds to the lower part of the corresponding Eurocode reliability class (RC2). For the next South African reliability class (RC3, typically buildings of five to fifteen storeys) $_T = 3.5$ approaches that of the undifferentiated Eurocode RC2.

Implicit Reliability Levels of Acceptable Practice

Keeping the design methodology compatible with the existing experience base is consistent with the evolutionary nature of codes and standards that require changes to be made cautiously and deliberately (Phoon 1995). In the spirit of a Bayesian approach towards reliability, proven experience is an important source of information that can be combined with other sources of data on variability and uncertainty in reliability-based design.

Accordingly this paper investigates the level of reliability of pile foundations designed in accordance with the static formula. This is done by determining the implicit levels of reliability for the current working stress design (WSD) methods for piles by comparing design values to reliability models for pile resistance. Reliability modelling of pile resistance is based on the uncertainty of pile resistance, as observed by the comparison of the interpreted resistance from pile tests and the predicted value for an extensive survey of pile tests done across southern Africa, representing a wide range of conditions, pile construction practices and configurations.

CONCEPTS OF RELIABILITY ANALYSIS AND CALIBRATION

Although reliability calibration and the analysis of existing practice form two distinct components of the application of reliability theory in design, they are so closely related that some concepts of their treatment in practice can share a common formulation. The common concepts are related to a specific level of reliability over a defined range of conditions. The following issues are relevant to reliability calibration of design procedure such as partial factor limit states design; therefore by implication also to reliability assessment of existing practice:

The representative level of reliability is either the target reliability in the case of calibration, or the implicit reliability in the case of assessing acceptable existing practice; conventionally expressed in terms of a reliability index as β_T and β_I respectively. The following alternative approaches apply to the representative reliability:

- An average value is taken across the range of conditions, although the value may be significantly exceeded in some cases - this is the view generally taken in Eurocode, also associated with relatively high levels of reliability (typically $\beta_T = 3.8$).
- A lower limit value is taken as a constraint, generally to be exceeded - this view is taken in South Africa, where the representative value is also relatively low (typically $\beta_T = 3.0$).

Consistency of reliability over the range of application is an objective to ensure that significantly different levels do not occur under different design conditions or cases; in particular systematically as a function of classes of applications (for example construction and/or soil type in the case of piles) or other design parameters. The following effects need to be considered:

- Conditions under which the lowest level of reliability is achieved will control the measures taken.
- Systematic exceeding of the representative reliability represents conditions which may be unjustifiably conservative.
- Consistency of reliability can be assessed in terms of the absence of different levels and the absence of trends, or at least smooth transitions related to continuous design parameters.

The level of confidence of calibration or assessment should take into account that it is at best an approximate process, due to the predictive nature of design. It is based on limited information, either for parametric calibration or assessment of existing practice such as presented here, or on the actual conditions in the case of design of a specific project. Calibration or assessment should therefore be moderated on the following basis:

- A limited level of confidence applies to both the required reliability levels (target or implied) and measures to achieve these - all based on acceptable performance of present practice.
- Best estimates of reliability is therefore generally acceptable, only reverting to conservative modelling when there is specific justification for such measures.

It should be noted that calibration back to existing practice does not imply maintaining the status quo just in a more complex format! With calibration, allowance can subsequently be made for rectifying conditions where reliability is inconsistent with the (present) general practice, either insufficient or unjustifiably conservative. Where insufficient reliability derives from uncertainty, as opposed to variability, appropriate measures can be considered, such as additional investigation consisting of gathering of data and improved modelling.

RELIABILITY MODELLING OF PILE RESISTANCE

The two predominant classes of uncertainty for geotechnical design can be distinguished as (i) uncertainties associated with design soil properties and (ii) calculation model uncertainties. With regard to geotechnical property uncertainties, significant research has been done to generate statistics on individual components of soil parameter uncertainty. Conversely, model statistics are relatively scarcer. In fact, the lack of model statistics is considered to be a key impediment to the development of geotechnical reliability-based design (Phoon 2005). This consideration provided the motivation for the investigation of model uncertainty of pile resistance as reported in the accompanying paper and by Dithinde et al (2011).

Model uncertainty, as defined for example in ISO 2394:1998, EN 1990:2002 and JCSS PMC (2001), reflects uncertainties of the structural mechanics model. Variability of variables, mainly actions, material properties and geometry is represented explicitly as basic variables in the performance function, as defined in Equation 1. In experimental determination of model uncertainty, values of basic variables are determined deterministically through testing. The implication is that model uncertainty represents not only the effect of the structural mechanics model, but also of all the sources of uncertainty, and even variability that is not explicitly captured in the testing process.

The modelling uncertainty reported in the accompanying paper not only reflects the uncertainty of the static pile design formula, but also the interpretation of site investigations and conversion of measurements into material properties. Due to the uncertainty of material properties and the absence of its representation as basic variables, the model factor can be considered to represent a probability model of pile resistance as predicted by the static pile formula. The procedure for using soil properties based on subsurface data surveys to predict pile resistance as described by Dithinde (2007) implies that the uncertainties from soil properties are incorporated in the predicted values. The pile resistance probability model can therefore be taken from the distributions and summary statistics presented in the accompanying paper.

In applying working stress design (WSD) procedures through the static pile design function, the emphasis is predominantly placed on pile resistance. Whilst loads are treated at nominal un-factored values, safety is treated by the application of a factor of safety to pile resistance. The initial parametric investigation of pile design practice is therefore based on considering pile resistance only. The effect of considering variability of loading is then considered subsequently.

Pile Resistance Only

In the definition of model uncertainty given in the accompanying paper, given here as Equation 4, the interpreted pile capacity (Q_i) can be taken to represent the probability representation of the pile resistance(R_{Rel}), and the predicted capacity (Q_p) the deterministic nominal pile resistance (R_n). R_{Rel} can therefore be expressed by Equation 5. The static pile design function is given by Equation 6 in terms of a factor of safety (FS) and nominal dead load (D_n) and live load (L_n). From Equations 5 and 6, the specific performance function given by Equation 7 can be converted into a parametric limit state function as shown in Equation 8. The implicit reliability index value ($_l$) an then be obtained from Equation 9 in terms of the probability model for M and the factor of safety FS which has a deterministic value.

$$M = \frac{Q_i}{Q_p} \tag{4}$$

$$R_{Rel} = M.R_n \tag{5}$$

$$\frac{R_n}{FS} = D_n + L_n \tag{6}$$

$$g = R_{Rel} - (D_n + L_n) \tag{7}$$

$$g = M \times R_n - \frac{R_n}{FS} = 0 = M - \frac{1}{FS} \tag{8}$$

$$\beta_I = \Phi^{-1}\left[P_f(M < \tfrac{1}{FS})\right] \tag{9}$$

Values for β_I can be obtained in terms of the pile classes identified in the accompanying paper. This is done by applying the reported statistics as parameter estimates for probability models for M. Comparison of β_I-values for alternative pile classes provides an indication of the representativeness and consistency of implicit reliability across the range of conditions represented by the M statistics.

As a point of departure the case of a single combined pile class (ALL) is used as a representative case to estimate $\beta_{I,Rep}$. This case is then used to investigate the influence of the probability distribution on β_I-values. The influence of pile class on β_I-estimates is considered below.

The estimate for $\beta_{I,Rep}$ is based on the lognormal distribution as standardised practice for resistance. Generally an overall factor of safety of 2.5 is regarded as an acceptable value for piles and has become a norm in southern Africa (Byrne & Berry 2008). As indicated in the accompanying paper, the normal distribution, which is conventionally used as the default first approximation distribution in reliability analysis, could also be considered. The mild degree of skewness indicated from the sample statistics presented in the accompanying paper can be taken into account by considering the general lognormal distribution. The results are summarised in Table 1, where values for the estimated distribution parameters are also given.

Table 1: Representative implicit reliability $_{I,Rep}$ for alternative probability distributions for combined pile class (ALL) and FS = 2.5

Distribution parameters		Distribution	Indicator	fit
Mean	1.10	Lognormal	Representative	3.52
Standard deviation	0.31	Normal	—	2.26
Skewness	0.24	General Lognormal	Low	2.45

The value for $_{I,Rep}$ is clearly sensitive to the distribution applied to represent M and therefore needs some interpretation: The value of $_{I,Rep}$ = 3.5 as obtained from the lognormal distribution is taken as an estimate of acceptable practice in accordance with standardised reliability procedures. The value of $_{I,Rep}$ = 2.3 obtained from the normal distribution is considered to be too low to reflect acceptable practice. The low value of skewness taken into account by the general lognormal distribution provides a slight improvement on this apparently low level of reliability; this result should be considered as a lower limit estimate of implicit reliability, as $_{I,Low}$ = 2.4.

An indication of the representativeness of these values of $_I$ across the range of pile classes is presented in Table 2. The value for $_{I,Rep}$ = 3.5 listed in Table 1 for the combined group (ALL) generally lies in the lower range of the values obtained for the various pile classes as listed in Table 2. The value of 3.5 is therefore taken to be in agreement with the approach of defining target reliability at a lower constraint value, and is thus ranked to indicate the mid-range value of implicit reliability.

Table 2: Range of implicit reliability values $_I$ and associated pile classes (FS = 2.5)

Range	Pile class	Lognormal ($\beta_{I,Rep}$)
Special	Driven piles in non-cohesive soil (D-NC)	3.1
Low	Non-cohesive soil (NC)	3.2

Mid	Combined group (ALL)	3.5
Mid +	Driven piles (D)	3.7
	Bored piles (B)	3.8
	Bored piles in non-cohesive soil (B-NC)	3.75
High	Driven piles in cohesive soil (D-C)	4.1
	Cohesive soil (C)	4.2
	Bored piles in cohesive soil (B-C)	4.3

The class of piles driven in non-cohesive soil (D-NC) is ranked at a special-range due to its low value in comparison to the representative implicit reliability. The more general pile class of non-cohesive soil (NC) is classified to be in the low-range. For all other pile classes, higher values for $_{I,Rep}$ are obtained (Mid+); with significantly higher values (High) obtained for bored piles in cohesive soils (B-C), as listed in Table 2.

R is therefore concluded that the values for $_{I,Rep}$ and $_{I,Low}$ obtained from Table 1 provide a reasonable representation of the implicit reliability of existing practice, but that the special case of driven piles in non-cohesive soils (D-NC) should be considered separately for systematically lower levels of reliability.

In addition to obtaining a lower limit estimate of the implicit reliability as based on the probability distribution, confidence level estimates of the distribution parameters can be applied. For this purpose the confidence limit estimates presented in the summary statistics in the accompanying paper are utilised. A single-sided 75% confidence limit estimate is used, as suggested by Eurocode EN 1990:2002 for cases where parameter estimation is based on vague prior distributions. The lower confidence limit value is used for the mean and the upper limit for the standard deviation. The confidence limit estimates $_{I,CONF}$ are listed in Table 3 for the two cases identified above as representative (ALL) and the special lower pile class (D-NC). Values are based on the confidence level distribution parameters, also listed in Table 3, applied to the lognormal distribution.

Table 3: Confidence limit ($_{I,Conf}$) values of implicit reliability as based on log-normal distribution and listed parameters (FS = 2.5)

	Combined group (ALL)	**Driven, non-cohesive (D-NC)**
I,Conf	3.2	2.3
Confidence level distribution parameters		
Mean	1.07	1.03
Standard deviation	0.32	0.40

Although the confidence limit implicit reliability listed in Table 3 is reduced for the representative case of the combined pile conditions listed in Table 1, the change from 3.5 to 3.2 is not too significant. For the special case D-NC the change from 3.3 to 2.2 is, however, indicative of not only the tendency of systematic lower value of implicit reliability, but also of the poor quality of its prediction.

The influence of the FS-value selected in pile design on $_{I}$-estimates is shown in Table 4. For comparison, target levels of reliability ($_{T}$) are also listed, as given by SANS 10160-1:2011 for different reliability classes of building structures. The comparison indicates reasonable agreement between $_{I, Low}$ for FS (2.0, 0.5 and 3.0) and for reliability classes (RC1, RC2 and RC3}. The values for $_{I, Rep}$ generally exceed that for the corresponding showing a trend of widening of the difference for the higher FS values and reliability classes.

Table 4: Implicit reliability ($_{I}$) as function of the selected value for FS, as compared to target reliability ($_{T}$) for reliability classes

	FS = 2.0	FS = 2.5	FS = 3.0
Combined group (ALL) ($_{I, Rep}$)	2.7	3.5	4.9
Driven, non-cohesive (D-NC) ($_{I, Low}$)	2.4	3.1	3.6
SANS 10160-1 Reliability Class	RC1	RC2	RC3
Target reliability ($_{T}$)	2.5	3.0	3.5

Implicit Reliability Based on Resistance and Loads

Expression of the performance function given by Equation 1 in terms of probabilistic models for resistance (R), dead (D) and live (L) is given in Equation 10.

$$g(R,D,L) = R - (D + L) \tag{10}$$

A normalised reliability model for Equation 10 can be obtained for parametric reliability analysis by representing each basic variable (X) by the ratio of mean to nominal value μ_X/X_n) and the relationship between the nominal values (R_n, D_n and L_n) given by the static pile design function (Equation 6). Similar to the treatment above, the resistance R is represented by the probability model for M as given by Equation 5. The load models reported by Kemp et al (1987) used for the conversion of structural design in South Africa from working stress to limit states design procedures listed in Table 5 can be used for models of D and L.

Table 5: Load models for reliability calibration (Kemp et al 1987)

Type of load	Code	Mean load / Nominal load	Coefficient of variation	Type of distribution
Dead (permanent) load	ANSI A58	1.05	0.10	Normal
	Australian	1.05	0.10	Lognormal
	SABS 0160	1.05	0.10	Lognormal
Live (office): lifetime max	ANSI A58	1.0	0.25	Gumbel
	Australian	0.7	0.26	Gumbel
	SABS 0160	0.96	0.25	Gumbel

Different loading conditions can be treated parametrically through the ratio L_n/D_n. A typical range of L_n/D_n ratios is 0.51.5 for concrete structures and 1-2 for steel structures (Melchers 1999). For foundations dead load would dominate, tending towards the lower range of load ratios. Based on this information, a practical range of L_n/D_n ratio of 0.5 to 2 was adopted as sufficiently representative of structures in general. The special cases of dead and live loads only are indicative of the outer limits of load conditions. For this reason the range of analysis was done for L_n/D_n between 0 and 2; the case for L_n only was also calculated. Parametric reliability analysis of Equation 10 was done using Second Order Reliability Method (SORM) software provided by Holický (2009).

The results for the representative reliability analysis ($_{\beta I, Rep}$) based on the model for the complete dataset (ALL) are shown in Figure 1(a); the results for the special case of driven piles in non-cohesive soil $_{\beta I, d-nc}$) are shown in Figure 1(b). Separate graphs are provided for the values of FS (2.0, 2.5 and 3.0). The results for the analysis of the complete version of Equation 10 are labelled as M, D, L (FS); the off-scale case of live load only is indicated as an arrow ($\rightarrow$) labelled M, L (FS); the results from the previous analysis considering pile resistance only are indicated as the horizontal line labelled M (FS).

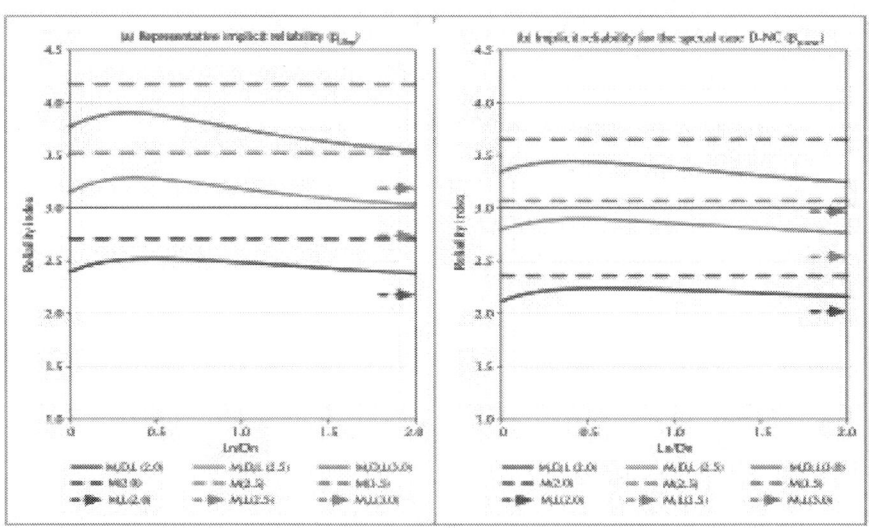

Figure 1: Implicit reliability of pile design including loading.

As can be expected, the inclusion of the effects of loading reduces the level of implicit reliability. Furthermore, the effects are dependent on the ratio of live to dead load (Ln/Dn), with trends similar to that obtained with load calibration analyses (Holický & Retief 2005). The lower values of implicit reliability occur for conditions dominated by live load, which generally can only be expected under exceptional conditions for pile foundations. Over the operational conditions of loading dominated by Dn the values for $_{I, Rep}$ compare well with the target reliability index values   listed for the various reliability classes listed in Table 4. For the special case of driven piles in non-cohesive soils, the values obtained for $_{I d-nc}$ are systematically lower than the corresponding values for $_T$.

DISCUSSION AND CONCLUSIONS

The resistance statistics for pile design practice in southern Africa, as reported in the accompanying paper, are applied in this paper to assess the implicit levels of reliability of such practice. This is done by deriving reliability models for pile resistance by applying the model statistics as parameter estimates to internationally standardised probability distributions for geotechnical and structural resistance. Values for implicit reliability, as expressed by the reliability index $_I$), are determined to obtain a measure of the representative level of performance of pile design practice. In addition to obtaining best estimate values for $_I$, conservative estimates are also made in terms of more severe interpretation of parameter estimates or probability distributions for pile resistance. Consistency of reliability is also investigated across the range of pile construction practice.

Implicit levels of reliability are derived for two reliability performance models:

i. Considering pile resistance only and neglecting loading as basic variable, in accordance with design practice reflected by the working stress design format for pile design, where a single factor of safety is applied to pile resistance.

ii. Including reliability models for dead (permanent) and live (variable) loads into the performance function, using the models on which the implementation and calibration of limit states design for South Africa were based.

The two main issues of concern for determining model statistics and applying these to reliability models for pile resistance identified in the accompanying paper are (i) the probability distribution used, and (ii) the scope of application as based on differentiated classes of pile conditions. It was found that the different plausible probability distributions have a more significant influence on the levels of implied reliability than differentiation into classes of pile conditions.

The default distribution applied in reliability analysis is generally the normal distribution, to represent the basic step from deterministic design practice to at least provide for the dispersion of basic variables. The lognormal distribution function at the same basic level of approximation has the added utility of not predicting negative values. This is particularly relevant when the lower tail of the distribution is considered, such as for resistance. However, values for β_I vary from a low value $\beta_{I,n} = 2.3$ for the normal distribution to a relatively high value $\beta_I = 3.5$ for the lognormal distribution, in both cases for the combined set of pile conditions and general design practice based on FS = 2.5. When skewness obtained from the model statistics is taken into account by applying the general lognormal distribution, $\beta_I = 2.4$ is obtained.

Selecting the lognormal distribution as basis to obtain a representative value for the reliability index $\beta_{I,Rep} = 3.5$ is based on consideration of standardised practice for reliability analysis, supported by the marginal preference obtained from the model statistics results presented in the accompanying paper. In accordance with Equation 3 the reliability index value corresponds with a probability of failure P_f of 2.10^{-4}.

A lower estimate $\beta_{I,Low} = 2.4$ ($P_f = 8.10^{-3}$) is based on the general lognormal distribution. Another lower limit estimate of β_I is based on the 75% confidence limit estimates of the distribution parameters, obtaining a value of $\beta_{I,CL} = 3.2$ ($P_f = 7.10^{-4}$).

Comparing the values for β_I for the combined set of pile conditions to the various pile classes, the following observations can be made: $\beta_{I,Rep} = 3.5$ provides a lower limit estimate value for β_I values for other pile classes based on construction method and/or soil type generally provide higher values. The exception is the case for driven piles in non-cohesive soil, where $\beta_{I,d-nc} = 3.1$ ($P_f = 1.10^{-4}$) is obtained; alternatively for non-cohesive soil $\beta_{I,NC} = 3.2$. When confidence level estimates are made, it is shown that the confidence limit value for driven piles in non-cohesive soil is as low as $\beta_{I,Conf} = 2.2$ ($P_f = 1.10^{-2}$).

When loads are also modelled as basic variables, the values for $\beta_{I,Rep}$ is somewhat reduced in the typical range of ratios for live to dead load to be expected for pile foundations. Values above 3.2 are obtained for live load less than dead load. The limited reduction in $\beta_{I,Rep}$ is indicative of the fact that reliability is dominated by the influence of pile resistance reliability. The reduction in β_I values for situations dominated by live load is indicative of the increasing role of the reliability of live load, which should optimally be provided for in terms of partial load factors, rather than being of concern for pile design reliability as such.

Comparison of values for $\beta_{I,Rep}$ corresponding to commonly adopted values for factor of safety FS (i.e. 2 - 3) generally shows good agreement with the target reliability β_T set in SANS 10160-1:2011 for different reliability classes. The values for implicit reliability for the three values of FS for dead load dominating conditions $\beta_{I, Rep}$ (2.5, 3.2 and 3.7) compares well with target reliability for the first three reliability classes β_T (2.5, 3.0 and 3.5). Nonetheless, the range of $\beta_{I, Rep}$ values obtained seem to be on the higher side for single piles, suggesting that current practice is conservative.

The reliability assessment of pile design practice does not only provide insight into the sufficiency of existing practice, but could also form the basis for achieving appropriate performance levels through reliability calibrated procedures. The rational basis for reliability calibration provided in SANS 10160-1:2011 can be applied in accordance with geotechnical limit states design procedures presented in SANS 10160-5:2011.

LIST OF NOTATIONS FOR RELIABILITY INDEX (β)

- β Reliability index, as related to prob- ability of failure given by Equation 3
- β_T Target level of reliability obtained through calibration of design expression
- β_I Reliability level implicitly achieved by existing practice, expressed in terms of the reliability index
- $\beta_{I, Rep}$ Indicative level of reliability taken to be representative of the set of pile conditions under consideration, usually considering pile design in general

- $_{I, Low}$ Lower limit estimates of implicit reliability based either on the selection of the probability distribution or the pile class
- $_{I, Conf}$ Reliability index value based on confidence limit estimates of distribution parameters
- $_{I, d-nc}$ Implicit reliability index value for the special case of driven piles in non-cohesive soil

REFERENCES

1. Baecher, G B & Christian, T 2003. Reliability and Statistics in Geotechnical Engineering. New York: Wiley.

2. Byrne, G & Berry, A D 2008. A Guide to Practical Geotechnical Engineering in Southern Africa, 4th edition. Germiston: Franki Africa (Pty) Ltd.

3. CE (Commision Européenne) 2002. Guidance Paper L: Application and Use of Eurocodes. Document CONSTRUCT 01/483 Rev.1, Brussels: CE.

4. Christian, J T 2004. Geotechnical engineering reliability: How well do we know what we are doing? The 39th Terzaghi Lecture. Journal of Geotechnical Engineering and Geoenvironmental Engineering, 130(10): 985-1003.

5. Day, P W & Retief, J V 2009. Provision for geotechnical design in SANS 10160, Chapter 5-1. In: Retief, J V & Dunaiski, P E (Eds), Background to SANS 10160. Stellenbosch: SUN MeDIA Press.

6. Dithinde, M 2007. Characterisation of model uncertainty for reliability-based design of pile foundations. PhD Thesis, Stellenbosch, South Africa: University of Stellenbosch.

7. Dithinde, M, Phoon, K K, De Wet M & Retief, J V 2011. Characterisation of model uncertainty in the static pile design formula. Journal of Geotechnical and Geoenvironmental Engineering, ASCE, 137(1): 333-342.

8. Dithinde, M & Retief J V 2013. Pile design practice in southern Africa. Part I: Resistance statistics. SAICE Journal, 55(1): 60-71.

9. European Committee for Standardization 1990. EN 1990:2002. Eurocode: Basis of Structural Design. Brussels: European Committee for Standardization (CEN).

10. European Committee for Standardization 1997. EN 1997:2004. Eurocode 7: Geotechnical Design. Part 1: General Rules. Brussels: European Committee for Standardization (CEN).

11. FHWA (US Federal Highway Administration) 2001. Load and Resistance Factor Design (LRFD) for Highway Bridge Substructures. Publication No.

12. FHWA-HI-98-032, Washington DC: FHWA. Holický, M & Retief J V 2005. Reliability assessment of alternative Eurocode and South African load combination schemes for structural design. SAICE Journal, 47(1): 15-20.

13. Holický, M 2009. Reliability Analysis for Structural Design. Stellenbosch: SUN MeDIA Press.

14. Holický, M, Retief, J V & Dunaiski, P E 2007. The reliability basis of design for structural resistance. Proceedings, 3rd International Conference on Structural Engineering, Mechanics and Computation (SEMC 2007), Cape Town, South Africa: Millpress, 1735-1740.

15. JCSS (Joint Committee on Structural Safety) PMC 200). Probabilistic model code. JCSS Working Materials [Online] http://www.jcss.ethz.ch (retrieved 1 Feb. 2011).

16. Kemp, A R, Milford, R V & Laurie, J P A 1987. Proposal for a comprehensive limit states formulation for South African structural codes. The Civil Engineer in South Africa, 29(9): 351-360.

17. Melchers, R E 1999. Structural Reliability: Analysis and Prediction. Chichester, New York: Wiley.

18. Milford, R V 1988. Target safety and SABS 0160 load factors. The Civil Engineer in South Africa, 30(10): 475-481.

19. Orr, T L L, Matsui, K & Day, P W 2002. Survey of geo-technical investigation methods and determination of parameter values. In: Honjo, Y (Ed), Foundation Design Codes and Soil Investigation in View of International Harmonization and Performance-Based Design. Proceedings, International Workshop on Foundation Design Codes and Performance-Based Design, Tokyo, Japan, 10-12 April, Lisse, Netherlands: AA Balkema.

20. Phoon, K K 1995. Reliability-based design of foundations for transmission line structures. PhD Thesis, Ithaca, NY, US: Cornell University.

21. Phoon, K K 2005. Reliability-based design incorporating model uncertainties. Proceedings, 3rd International Conference on Geotechnical Engineering combined with the 9th Yearly Meeting of the Indonesian Society for Geotechnical Engineering, 191-203.

22. Rackwitz, R 2008. Optimization with a LQI acceptance criterion, Annexure 5. In: JCSS Risk Assessment in Engineering - Principles, Systems Representation. pp 78-79. [Online]http://www.jcss.ethz.ch/publications/background/Risk_background   Doc_LQI_Optimization.pdf (retrieved on 1 Feb. 2011).

23. Retief, J V & Dunaiski, P E 2009. The limit states basis of structural design for SANS 10160-1, Chapter 1-2. In: Retief, J V & Dunaiski, P E (Eds), Background to SANS 10160.

24. Stellenbosch: SUN MeDIA. SABS 1972. SABS 088:1972. Code of Practice for Pile Foundations. Pretoria: South African Bureau of Standards.

25. SAKO (Joint Committee of NKB and INSTA-B) 1999. Basis of Design of Structures. Proposal for Modification of Partial Safety Factors in Eurocodes. Oslo, Norway: NKB Committee and Work Reports.

26. SANS 2004. SANS 2394:2004. General Principles on Reliability for Structures. Adoption of International Standard ISO 2394-1998, Pretoria: South African Bureau of Standards.

27. SANS 2011a. SANS 10160-1:2011. Basis of Structural Design and Actions for Buildings and Industrial Structures. Part 1: Basis of Structural Design. Pretoria: South African Bureau of Standards.

28. SANS 2011b. SANS 10160-5:2011. Basis of Structural Design and Actions for Buildings and Industrial Structures. Part 5: Basis for Geotechnical Design and Actions. Pretoria: South African Bureau of Standards.

Chapter 3

Evaluation of the Ultimate Capacity of Friction Piles

Wael N. Abd Elsamee

Faculty of Engineering, Sinai University, El Arish, Egypt

ABSTRACT

The precise prediction of maximum load carrying capacity of bored piles is a complex problem because the load is a function of a large number of factors. These factors include method of boring, method of concreting, quality of concrete, expertise of the construction staff, the ground conditions and the pile geometry. To ascertain the field performance and estimate load carrying capacities of piles, in-situ pile load tests are conducted. Due to practical and time constraints, it is not possible to load the pile up-to failure. In this study, field pile load test data is analyzed to estimate the ultimate load for friction piles. The analysis is based on three pile load test results. The tests are conducted at the site of The Cultural and Recreational Complex project in Port Said, Egypt. Three pile load tests are performed on bored piles of 900 mm diameter and 50 m length. Geotechnical investigations at the site are carried out to a maximum depth of 60 m. Ultimate capacities of piles are determined according to different methods including Egyptian Code of practice (2005), Tangent-tangent, Hansen (1963), Chin (1970), Ahmed and Pies (1997) and Decourt (1999). It was concluded that approximately 8% of the ultimate load is resisted by bearing at the base of the pile, and that up to 92% of the load is resisted by friction along the shaft. Based on a comparison of pile capacity predictions using different method, recommendations are made. A new method is proposed to calculate the ultimate capacity of the pile from pile load test data. The ultimate capacity of the bored piles predicted using the proposed method appears to be reliable and compares well to different available methods.

INTRODUCTION

Pile foundation is an important link in transferring the structural load to the bearing ground located at some depth below ground surface. The design of piles accounts for various parameters such as the nature of substrata, depth of ground water table, depth of the bearing stratum, and type and level of load to be supported. To ascertain the field performance and estimate the load carrying capacity, in-situ pile load tests are relied upon.

A simple method for calculating static shaft resistance of a pile driven into clay is presented by Mirza (1997) [1]. The method is based on correlations derived for marine clays between index properties and strengths. Applications of the method to half a dozen full scale pile load tests of high quality are described. Except for short piles in very stiff to hard clays, the predictions agree well with the field test measurements. The correlation presented allows an assessment of residual skin friction and indicates the importance of the liquidity index of the clay in static capacity calculations.

Dewaikar and Pallavi (2000) presented analysis of field pile load tests data to estimate the ultimate pile load. The analysis is based on forty pile load tests results collected from various infrastructure and building sites in Mumbai region of India. Collected data is analyzed using various graphical and semi-empirical methods available in literature [2].

Nabil (2001) studied the behavior of bored pile groups in cemented sands by a field testing program at a site in South Surra, Kuwait. The program consisted of axial load tests on single bored piles in tension and compression. Two groups of piles, each consisting of five piles were tested. The spacing between the piles in the groups was twoand three-pile diameters. The calculated pile group efficiencies were 1.22 and 1.93 for a pile spacing of twoand three-pile diameters, respectively. Since settlement usually controls the design of pile groups in sand, the group factor, defined as the ratio of the settlement of the group to the settlement of a single pile at comparable loads in the elastic range, was determined from test results [3].

Abdelrahman et al. (2003) suggested that axial pile loading tests on single pile may offer the justification of the pile design load. Codes for deep foundations design stipulate the acceptance criteria for piles tested in compression based on specified limits for pile settlement at specified load levels. The researchers examined the different methods used in interpreting pile load test results. Sixty-four continuous flight auger piles were tested using the maintained load test method and the results were analyzed using the different methods of interpretation [4].

Wehnert and Vermeer (2004) analyzed the load results of short large diameter bored pile tested in Germany. The results for total resistance as well as for base and shaft resistance are presented. The pile is assumed to be linear elastic. Different constitutive models for the subsoil such as elastic-plastic, Mohr-Coulmb, are used [5].

A new approach for the design of large diameter bored piles resting on cohesionless soils was suggested by Radwan et al. (2007) [6]. The approach is based on the results obtained from finite element analysis performed using data from thirty case histories of large diameter bored piles collected from several construction projects. Both unit end bearing and skin friction resistance are estimated taking the settlement criterion into account. Mohr-Coulomb constitutive model is used in the numerical model. Eventually, statistical study is conducted to evaluate the improvement, accuracy, and reliability of design using the new approach, compared with the prediction of the Egyptian Code (2005) [7].

Akbar et al. (2008) presented the experience gained from four pile load tests at a site in the North West Frontier Province of Pakistan. Geotechnical investigations at the site are carried out to a maximum depth of 60 m. The soil at the site is predominantly hard clays within the investigated depth with thin layers of gravels and boulders below 40 m depth. Four piles of diameters varying from 660 mm to 760 mm and length ranging between 20 m and 47.5 m were subjected to axial loads. Using the pile load test results, back calculations are carried out to estimate the appropriate values of pile design parameters [8].

A probabilistic model as a complementary mathematical base for the traditional deterministic approach to quantify the selection of a factor of safety for each term of the load equation of friction piles in clay is presented by Al Jairry (2009) [9].

From the above, the variation in the load estimates of available methods is too much. Thus, additional study on friction pile capacity is needed to be done. However, the objective of this study is to provide the results of pile tests and develop a formula for closer prediction of the pile capacity.

SOIL INVESTIGATION

There have not been many tests on the soil in Port Said in Egypt. The investigated site is the Cultural and Recreation Complex project located in the city of Port Said. The project is built on an area of approximately 50 × 70 m. A comprehensive geotechnical investigation was conducted. The investigation included seven borings. The general layout of site is shown in Figure 1.

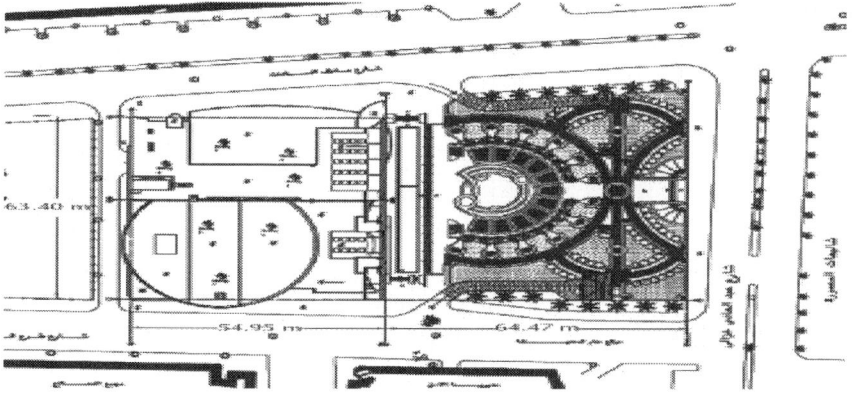

Figure 1: General layout of the site.

Soil Stratification

The soil profile in the investigated site is shown in Figure 2. The profile indicates that the following soil stratifications are encountered:

1) From elevation 0.00 to –10.00 m calcareous silt sand with broken shells.

2) From elevation –10.00 to –17.00 m soft silty-clay with interval of sand.

3) From elevation –17.00 to –49.00 m soft silty-clay with traces of sand.

4) From elevation –49.00 to –52.00 m calcareous sit sand.

5) From elevation –52.00 to –60.00 m hard silty-clay with intervening calcareous silty-sand.

The ground water table has been found to be at 0.70 meter from the ground surface.

PREDICTION OF PILE LOAD CAPACITY USING EGYPTIAN CODE

Various field and laboratory tests are carried out during the geotechnical investigation for the evaluation of subsurface conditions and the pile design parameters at the project site. The code pile capacities are calculated using the provisions of the Egyptian code (2005) [7]. The pile diameter is taken as 900 mm and pile length is 50 m. Tables 1-3 summarize the soil properties as well as outlining the calculated pile resistance (shaft friction and end bearing). Figure 3 shows the calculated ultimate capacity of the pile. Based on data from the figure, the ultimate pile capacity, Quit is obtained as 4622.81 kN/m². By applying a factor of safety, F.S. of 2, the allowable design pile capacity, Quall is 2311.41 kN/m². The allowable bearing capacity of the pile adopted for the design is taken as 2300 kN/m².

PILE LOAD TESTS

Three pile load tests are performed on bored piles of 900 mm diameters and 50 m lengths. One of the piles is non-working pile test #1 and two are working piles tests #2 and #3. The nonworking pile test #1 is loaded to twice the working load of 230 ton while the working piles for tests #2 and #3 are loaded to 1.5 times the working load.

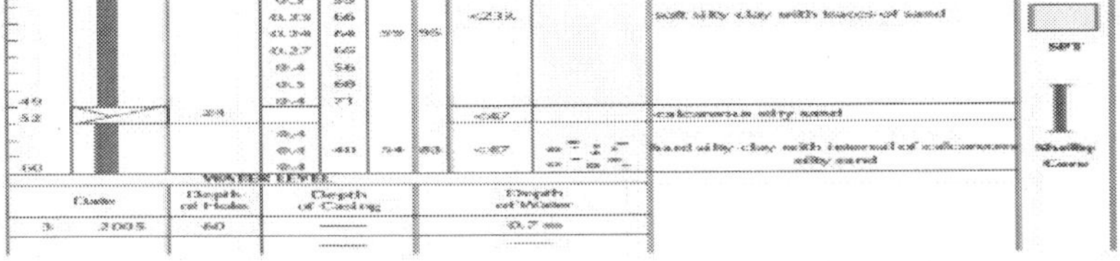

Figure 2: Soil profile of the investigated site.

Table 1: Calculated skin friction to be used in the design of pile according to the Egyptian Code [7]

	Layer depth under the SBL [lli	Soil type	Av. SPT N value	Undrained cohesion Cu [kN/m²)	Depth [m]	SPT	Layer thickness[m]	Skin friction τ[kN/m²]	Friction pile load Q [KN]
1	0- 5	CS-S	6			-	3	0	0.0
2	5 - 10	CS-S	24	-	2 - 7.5	20 - 30	5	75	1060.3
3	10 - 17	SS-C		20			7	20	395.8
4	17 - 49	HS-C		20	>7.5		32	20	1809.6
7	49 - 53	CS-S	>50		-	>50	2.1	100	593.8

Skin friction at settlement of 0.2 Sg = 0.9 cm, Qt = 3859.4; For Sg = 5%, D = 4.5 cm.

Table 2: Calculated end bearing resistance to be used in the design according to the Egyptian Code [7]

Point	Settlement [cm]		Bearing stress [1(N/na²]]	Pile area [m²]	End bearing pile load [KN]
0	0	0	0	0	0
A	0.2 Sg	1	500	0.64	318.09
B	0.3 Sg	1.35	700	0.64	445.32
C	Sg	4.5	1200	0.64	763.41

Table 3: Total pile load to be used in the design according to the Egyptian Code [7]

Point	End bearing pile load [KN]	Friction pile load Q KN]	Total pile resistance
O	0	0	0
A	318.09	3859.4	4177.49
B	445.32	3859.4	4304.72
C	763.41	3859.4	4622.81

Thus, the code ultimate capacity of pile = 3859.4 + 763.41 = 4622.81 kN/m²

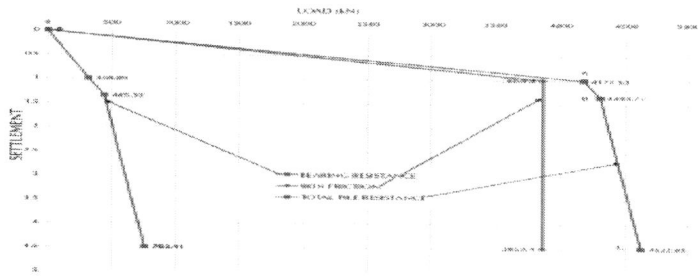

Figure 3: Shows the relationship between calculated capacity and settlement for the bored pile according the Egyptian Code.

Reaction System

The reaction system for the test piles was provided by a test head restrained by twelve ground anchors distributed around the pile as shown in the test setup in Figure 4.

Loading of Pile

The load was applied using three hydraulic jacks placed between the pile head and the anchored test head as shown in Figures 4 and 5. The loading cycle's increment adopted for the test piles according Egyptian code.

Test Measurements

Measurement of load the load was measured by calibrated load cells with digital readout device. Load cells were seated on top of spherical bearing plates placed above the hydraulic jacks. Also, the applied load was checked by recording the applied hydraulic pressure by a pressure gauge mounted on the pumping unit.

Measurement of pile head settlement Settlement of the pile head is measured using three dial gauges of precision of 0.01 mm.

Test Results

- General Observation during tests a) Settlement of pile did not reach 10% of its nominal diameter.
- The test piles did not show any sign of geotechnical failure. This means that the test piles did not continue to settle or sink without increase in the applied load.
- No section of the test piles failed structurally.

The load-settlement relationships for pile load tests are shown in Figure 6.

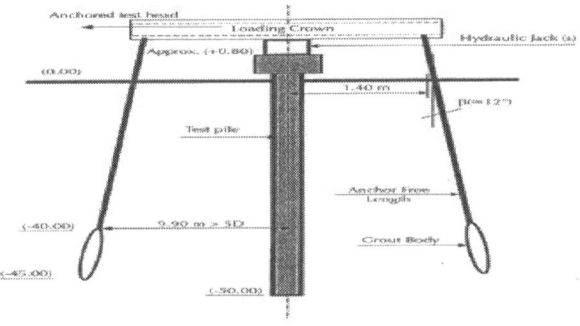

Figure 4: Test setup.

Figure 5: Pile loading setup.

- Head Settlement is recorded in Table 4. It is noted that no sign of plunging is detected.

ULTIMATE CAPACITY OF PILES

The ultimate capacities of the piles are determined from the load test results using different approaches.

Tangent—Tangent Method

Applying tangent—tangent method, a plot is made between load divided by cross sectional area of pile and the settlement on semi logarithmic scale as shown in Figure 7 for working pile load test #2 [7].

Hansen Method (1963)

Applying Hansen Method the square root of each settlement value from field load test data divided by the corresponding load value is plotted against the settlement as shown in Figure 8 for working pile load test #3. Estimation of the ultimate load by Hansen Method is given by the formula [10].

$$Q_u = \left(2C_1C_2\right)^{1/2} \tag{1}$$

Where:

Q_u = ultimate load capacity.

C_1 = slope of the best fitting straight line.

C_2 = y-intercept of the straight line.

Chin's Method (1970)

Applying Chin's method, a plot is made between settlement divided by corresponding load and the settlement as shown in Figure 9 for non-working test pile #1. The inverse slope of the straight line gives the ultimate load as proposed by Chin [11].

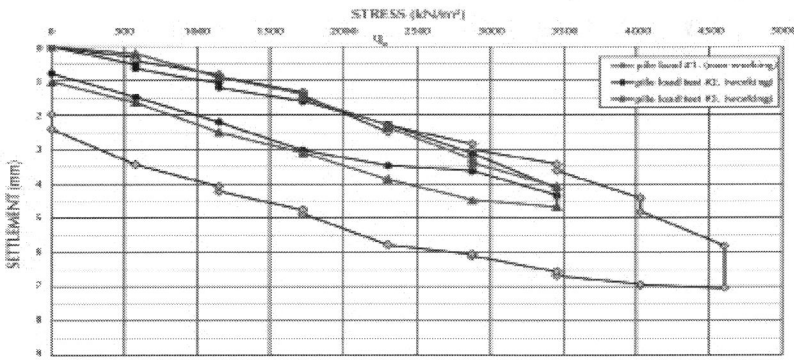

Figure 6: Load-settlement relationship for non-working pile load test #1.

Table 4: Recorded head settlement

Test No.	Pile #1 non-working	Pile #2 working	Pile #3 working
Settlement at 230 tons (anticipated working load)	2.27 mm	2.29 mm	3.40 nun
Settlement at 345 tons (150% of the working load)	3.62 mm	4.33 nun	3.87 nun
Settlement at 460 tons (200% of the working load)	7.03 mm		
Residual (parameter) settlement	1.97 nun	0.77 nun	1 03 mm

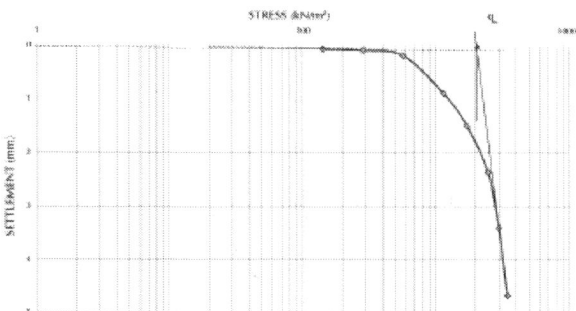

Figure 7: Ultimate pile capacities by tangent—tangent method for working pile load test #2.

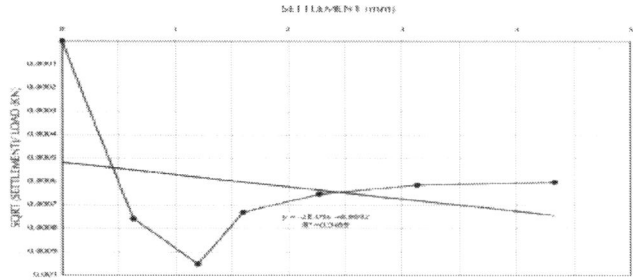

Figure 8: Ultimate pile capacities by Hansen method for working pile load test #3.

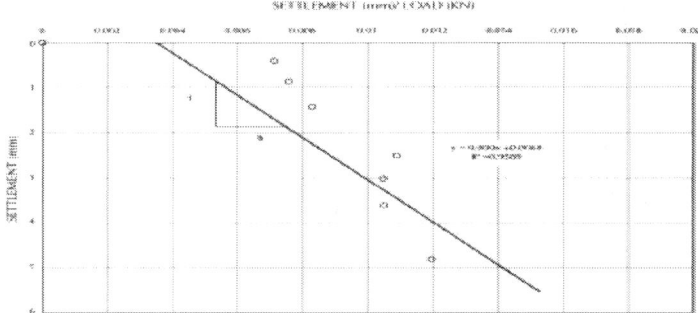

Figure 9: Ultimate pile capacity by Chin method for non-working test pile #1.

Ahmad and Pise (1997)

Ahmad and Pise (1997) proposed a reduction factor to Chin's extrapolated value of the ultimate capacity. In the settlement/load vs. settlement plot, it was observed that, generally two straight lines could be drawn through these points. As shown in Figure 10 for non-working test pile #1, the ratio of settlement S (settlement between the point of intersection of two straight lines and that corresponding to final test load) to S (total settlement) is taken to be the reduction factor (RF) for that set of test data [12]. However, reduction factor (RF) is given by the following

$$RF = \frac{\Delta S}{S} \qquad (2)$$

Where:

RF = Reduction factor.

Q_{mod} = Modified Chin's value of ultimate capacity.

Q_{ch} = Chin's value of ultimate capacity.

Decourt's Extrapolation (1999)

Applying Decourt's Extrapolation by dividing each load by its corresponding settlement and ploting the resulting values against the applied load. A linear regression over the apparent line (last three points) determines a line. Decourt identified the ultimate load as the intersection of this line with load axis as shown in Figures 11 for working test pile #3 [13].

PROPOSED METHOD FOR DETERMINATION OF ULTIMATE PILE CAPACITY FROM LOAD TEST

The load vs settlement behavior of the pile is extrapolated using an empirical method. The estimation of ultimate load consists of two steps as given below:

1)   Plotting load settlement curve from field load test data as shown in Figures 12-14.

2)   The ultimate pile capacity is given by the empirical formula:

$$Q_u = \left[\frac{1}{0.445my} \right] \tag{3}$$

Where:

Q_u = ultimate load capacity (kN).

m = slope of the trend straight line.

y = y-intercept of the straight line (as a value without sign).

COMPARISON BETWEEN DIFFERENT METHODS FOR DETERMINATION OF ULTIMATE PILE CAPACITY

The calculation of the ultimate capacity of piles and the corresponding factors of safety using the above mention methods are summarized in Table 5.

The ultimate loads obtained by various methods from the pile load test results are shown in Figure 15.

LOAD CARRIED BY END BEARING AND FRICTION ALONG SHAFT

From Table 6 the values of the ultimate pile capacity were taken to evaluate the percentage of friction and end bearing capacity from Figure 3. Based on the above findings, it was found that the percentage of load carried by friction along the pile shaft and the end bearing are shown in the following Table 6.

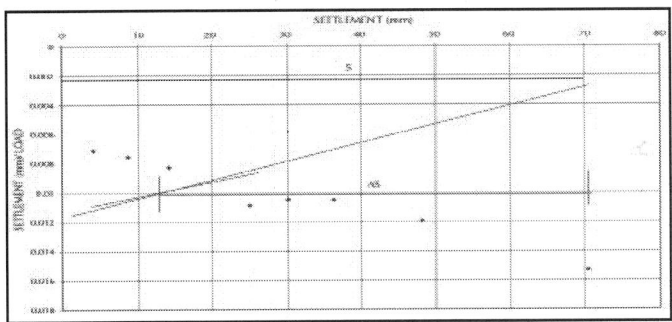

Figure 10: Ultimate pile capacity by Ahmad and Pise method for non-working test pile #1.

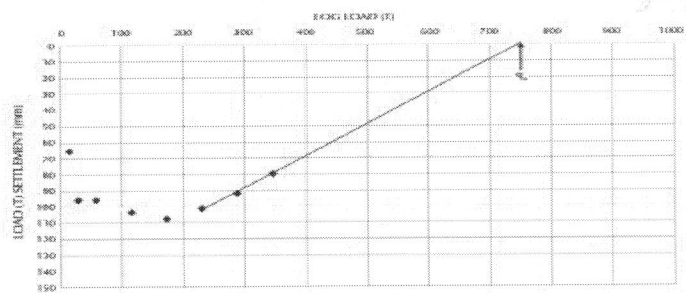

Figure 11: Ultimate pile capacities by Decourt's extrapolation method for working test pile #3.

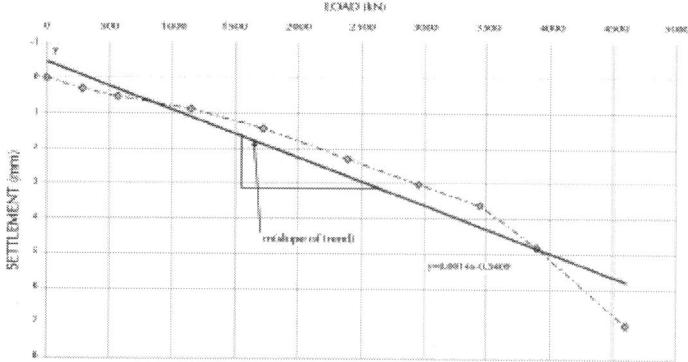

Figure 12: Ultimate pile capacity using proposed method for non-working test pile #1.

CONCLUSIONS

From the testing program and comparable study conducted, the following conclusions are arrived at:

1) The percentage of friction load carried by the shaft is approximately 85% to 90% and the percentage of load carried by the end bearing is 15% to 10%.

2) Hansen (1963) method gives higher values of ultimate capacity carried by the pile than the other methods.

3) A new proposed method to calculate the ultimate capacity of pile from pile load test is presented.

4) The proposed method for determining the ultimate capacity of friction piles appears to give results that are in good agreement with the analytical predictions.

5) The proposed method is good to apply, easier, quicker, more reliable, does not give max or min numbers as compared to some others.

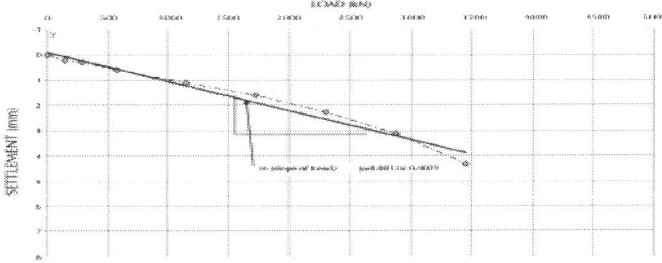

Figure 13: Ultimate pile capacity piles using proposed method for working test pile pile #2.

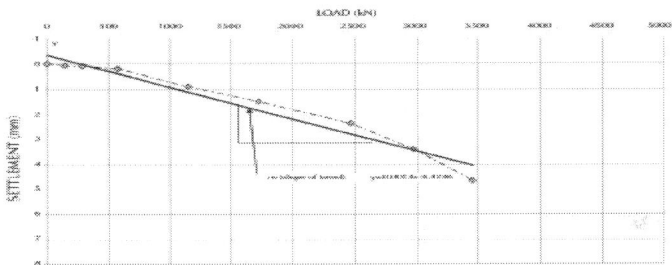

Figure 14: Ultimate pile capacity pile using proposed method for working test pile #3.

Table 5: Ultimate capacity and factor of safety (F.S.) of pile using different methods

Test No	Pile# 1 non workingworking		Pile #2 working		Pile #3 working	
Method	Q_{ult} (kN)	F.S.	Q_{ult}(kN)	F.S.	Q_{ult}(k N)	F.S.
Tangent	5600.00	2.43	5300.00	2.30	4400.00	2.00
Hansen (1963)	9128.71	3.97	5000.00	2.17	3227.49	1.40
Chin (1970)	8333.33	3.62	5555.56	2.14	4166.67	1.81
Aluned & Pise (1997)	6641.66	2.88	4381.58	1.91	3319.06	1.44
Decatut (1999)	6990.00	3.03	7300.00	3.17	5750.00	2.50
Presentstudy	4720.99	2.05	4658.36	2.03	4080.49	1.77

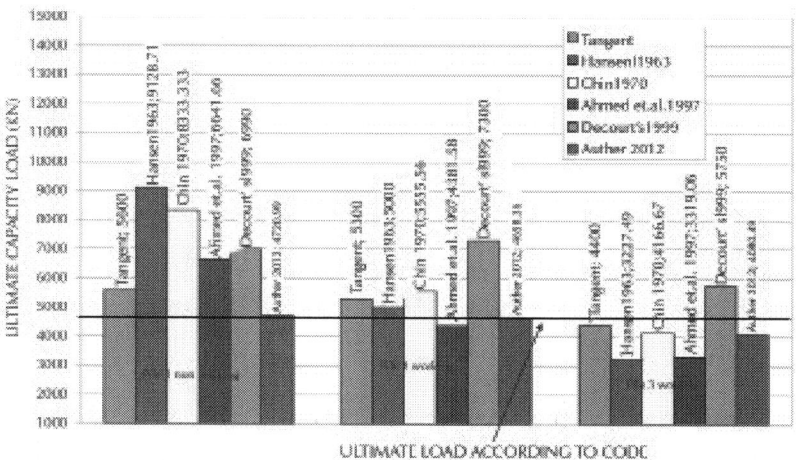

Figure 15: Comparison of ultimate pile loads using different methods.

Table 6: Percentage of ultimate load carried by end bearing and friction

Test Pile No	Pile #1 non-working		Pile #2 working		Pile #3 working	
Method	Skin friction %	End bearing %	Skin friction %	End bearing %	Skin friction %	End bearing %
Code load	90.8	9.20	92.5	7.50	92.2	7.80
Tangent	84.3	15.7	83.9	16.1	87.2	12.8
Hansen (1963)	88.1	11.9	90.8	9.20	88.1	11.9
Chin (1970)	86.8	13.2	83.9	16.1	83.7	16.3
Ahmed and Pise (1997)	85.0	15.0	83.5	16.5	86.0	14.0
Decourt (1999)	85.4	14.60	85.70	14.30	84.10	15.90
Present study	91.2	8.80	92.3	7.70	90.6	9.4

ACKNOWLEDGEMENTS

The author would like to acknowledge the Fetih Construction Company and Pauer-Egypt Company for their valuable assistance.

REFERENCES

1. U. A. A. Mirza, "Pile Skin Friction in Clays," International Journal of Offshore and Polar Engineering, Vol. 7, No. 1, 1997, pp. 538-540.

2. D. M. Dewaikar and M. J. Pallavi, "Analysis of Pile Load Tests Data," Journal of Southeast Asian Geotechnical Society, Vol. 6, No. 4, 2000, pp. 27-39.

3. F. I. Nabil, "Axial Load Tests on Bored Piles and Pile Groups in Cemented Sands," Journal of Geotechnical and Geoenvironmental Engineering, Vol. 127, No. 9, 2001, pp. 766-733.

4. G. E. Abdelrahman, E. M. Shaarawi and K. S. Abouzaid, "Interpretation of Axial Pile Load Test Results for Continuous Flight Auger Piles," Emerging Technologies in Structural Engineering, Proceedings of the 9th Arab Structural Engineering Conference, Abu Dhabi, 29 November- 1 December 2003, pp. 791-802.

5. M. Wehnert and P. A. Vermeer, "Numerical Analysis of Load Test on Bored Piles," Proceedings of the Ninth International Symposium on "Numerical Models in Geomechanics", Ottawa, 25-27 August 2004, pp. 1-6.

6. M. Radwan, A. H. Abdel-rahman, M. Rabie and M. F. Awad-Allah, "New Suggested Approach for Design of Large Diameter Bored Piles Based on Finite Element Analysis," Twelfth International Colloquium on Structural and Geotechnical Engineering (12th ICSGE), 10-12 December 2007, Cairo, pp. 340-357.

7. A Egyptian Code, "Soil Mechanics and Foundation," Organization, Cairo, 2005.

8. A Akbar, S. Khilji, S. B. Khan, M. S. Qureshi and M. Sattar, "Shaft Friction of Bored Piles in Hard Clay," Pakistan Journal of Engineering and Applied Science, Vol. 3, 2008, pp. 54-60.

9.   H. H. Al Jairry, "Exact Probability Equation for Friction Piles in Clay," Iraqi Journal of Civil Engineering, Vol. 6, No. 1, 2009, pp. 791-802.

10.   J. B. Hansen, "Discussion on Hyperbolic Stress-Strain Response, Cohesive Soils," Journal for Soil Mechanics and Foundation Engineering, Vol. 89, 1963, pp. 241- 242.

11.   F. K. Chin, "Estimation of the Ultimate Load of Piles from Tests Not Carried to Failure," Proceedings of Second Southeast Asian Conference on Soil Engineering, Singapore City, 11-15 June 1970, pp. 81-92.

12.   F. Ahmed and P. J. Pise, "Pile Load Test Data-Interpretation & Correlation Study," Indian Geotechnical Conference, Vadodara, 17-20 December 1997, pp. 443-446.

13.   L. Decourt, "Behavior of Foundations under Working Load Conditions," Proceedings of the 11th Pan-American Conference on Soil Mechanics and Geotechnical Engineering, Foz Dolguassu, August 1999, Vol. 4, pp. 453-488.

Chapter 4

Experimental Study of Dynamic Characteristics on Composite Foundation with CFG Long Pile and Rammed Cement-Soil Short Pile

Jihui Ding[1], Yanliang Cao[1], Weiyu Wang[2], Tuo Zhao[2], and Junhui Feng[3]

[1]College of Civil Engineering, Hebei University, Baoding, China
[2]Hebei Academy of Building Research, Shijiazhuang, China
[3]China Metallurgical Design and Research Institute Co., Ltd., Baoding, China

ABSTRACT

Based on the idea of optimization design of pile type, the two kinds of the typical pile type are selected, which containing flexibility pile (e.g. rammed cement-soil pile is for short RCSP), and rigid pile (e.g.

~ cement-flyash-gravel pile is for short CFGP). The three kinds of the composite foundation are designed, which are CFGP, CFG long pile and CFG short pile (for short CFGLP-CFGSP), CFG long short pile and rammed cement-soil short pile (for short CFGLP-RCSSP). Natural earthquake is simulated by using the engineering blasting; the dynamic characteristics and dynamic response of the composite foundation are studied through field test. CFGLP-RCSSP is closed to linear relation. The bearing capacity of the four composite foundation of the CFGP, CFGLP-CFGSP, and CFGLPRCSSP in the site are 225 kPa, 179 kPa, and 197 kPa, separately increases 150%, 98.8% and 119% compared to the natural foundation. The vibration main frequency is mainly depended on properties of foundation soil and piles between vibration source and measuring point, pilling load value. Horizontal vibration main frequency greater than the vertical vibration main frequency and the vertical vibration main frequency close to the first-order natural frequency of composite foundation. With the pilling load increasing, the CFGLP-RCSSP pile composite foundation combined frequency decreased. Under the same blast energy, the acceleration peak on the CFG pile composite foundation is less than CFGLP-CFGSP the corresponding values, as the load increases, the peak acceleration gently. CFG pile composite foundation is favorable on seismic. The distribution of peak acceleration is consistent within 4 m from pile top in the CFGLP_RCSSP composite foundation. The maximum of the horizontal acceleration peak along the pile body occurs at a distance of pile top 4 m or the pile top, and that of vertical acceleration peak occurred at a pile top.

INTRODUCTION

The composite foundation is that the part soil body in the natural ground foundation is reinforced or replaced during the ground treatment, and load is born by reinforced body and soil body around the pile [1]. Design theory of single pile composite foundation is relatively mature, and has certain limitations and shortcomings. The pile stiffness is smaller and the pile body has certain bond strength in the flexible pile composite foundation. The most commonly flexible piles are mixing cement soil pile [2] [3], rammed cement-soil pile [4] [5] pile, and so on. The strength of the flexible piles is low and load cannot be effectively transmitted to the lower part of the pile. When the top of the soil-cement pile is crushed, the side friction of the pile length range did not develop out.

The pile in the rigid pile (i.e. CFG pile) composite foundation has higher strength, with large adjustment range of bearing capacity of composite foundation [6]. Usually the rigid pile has happened with piercing failure, and the pile body material strength has not fully developed out. To give full play to the advantages of various types of pile, the composite foundation to form by different typed piles combined together [7] [8], can maximize the advantages of various types of pile. With the rapid development of economy, strength, length of pile in composite foundation can be greatly improved, and greatly improve the bearing capacity of composite foundation. The original design theory of composite foundation cannot meet the requirements, and dynamic problems of composite foundation have become the focus of attention. Study on the seismic performance of the composite foundation is the main application of numerical analysis and simulation of the composite foundation of finite element software [9] -[11] , this method still remain at the theoretical level, have not been applied to the actual design. Wang Weiyu, Zhao Tuo, Ding Jihui etc. studied dynamic characteristic and its influence factors of cement soil pile and CFG pile composite foundation under the action of the blasting vibration [12] -[15] .

Optimization is made to CFG pile and rammed cement-soil pile, and the composite foundations of CFG pile, CFG long pile and CFG short pile, CFG long pile and rammed cement-soil pile are, the designed. The stress and dynamic characteristics of the three composite foundation are studied through field tests.

THE INTRODUCTION OF THE TEST SITE

The test site is located in Shijiazhuang Heibei province. In the 20 m depth, the soil layers mainly are yellow silt clay, fine sand, middle sand and silt clay. In the 20 m driving depth, the underwater is not seen. There is not the harmful geologic action in the site. The main parameters of soil layer as shown in Table 1.

MODEL TEST AND SCHEME OF THE SITE

Three kinds of composite foundation model are designed: CFGP, CFGLP-CFGSP, CFGCP-RCSLP composite foundation. The model design parameter of composite foundation is shown in Table 2. CFG pile adopts C20 commercial concrete. Blasting is used as the vibration resource. The diameter of blasting hole is 50 mm. The Explosives are buried in the hole and then backfill tamping. The vibration is picked by acceleration sensors. The arrangement of the piles and measuring elements are shown in Figures 1-4.

The upper load is supplied by pilling concrete block, and each load of the composite foundation is added by an electric pressure pump-hydraulic jack. The square steel is 2.0 × 2.0 m as the loading plate.

THE ANALYSIS OF EXPERIMENTS RESULT

Load-Settlement Curves

Combining the three kinds load test of composite foundation, the load-settlement curves as shown in Figure 4. From Figure 4, compared with natural foundation, the bearing capacity of composite foundation of CFGP, CFGLP-CFGSP and CFGLP-RCSSP increases obviously and the deformation of composite foundation decreased compared to the Natural Foundation. The nonlinear degree of the p - s curves of combined pile decreases, and CFGLP-RCSSP is closed to linear relation. The bearing capacity of the four composite pile of the CFGP, CFGLP-CFGSP, and CFGLP-RCSSP in the site are 225 kPa, 179 kPa, and 197 kPa, separately increases 150%, 98.8% and 119% compared to the Natural Foundation.

Table 1: Mainly parameter of soil layer

No.	h_i/m	f_{ak}/kpa	E_s/Mpa	f_{sk}/kpa	f_{qk}/kpa
1)	0.5	130	6.17	60	
2)	1.5	140	10	55	800
3)	4.0	200	11.5	82	1500
4)		270	5.84	65	1000

Where h_i is thickness of the soil layer, f_{ak} is characteristic value of bearing capacity, E_s is compression Modulus, f_{sk} is ultimate shaft resistance f_{qk} is ultimate tip resistance.

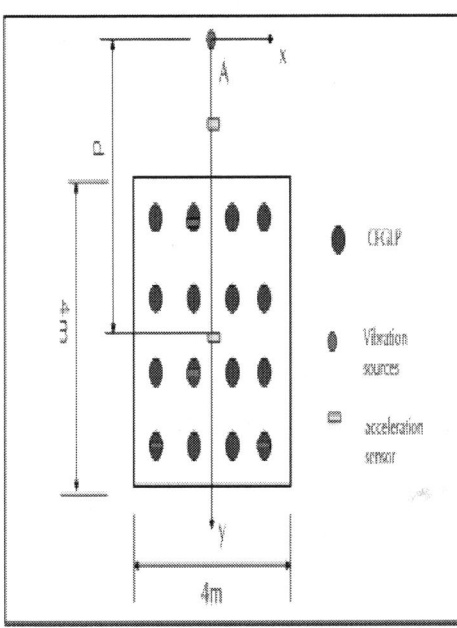

Figure 1: The arrangement of the piles and measuring elements of the Model 1.

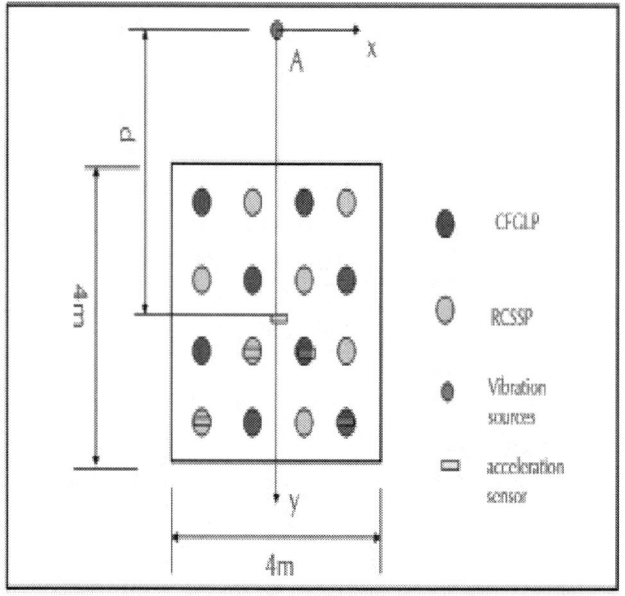

Figure 2: The arrangement of the piles and measuring elements of the Model 2.

Table 2: Model design parameter of composite foundation

Model	Type	Pile length/m	Pile diameter/ mm	Pile spacing/ mm	Replacement rate
1	CFGP	6.0	350	100	0.09616
2	CFGLP	6.0	350	200	0.04808
	CFGSP	4.0	350	200	0.04808
3	CFGLP	6.0	350	200	0.04808
	RCSSP	4.0	350	200	0.04808

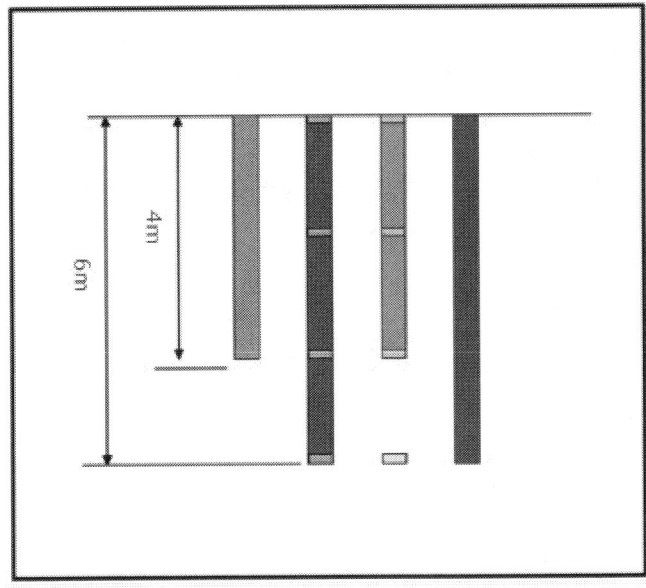

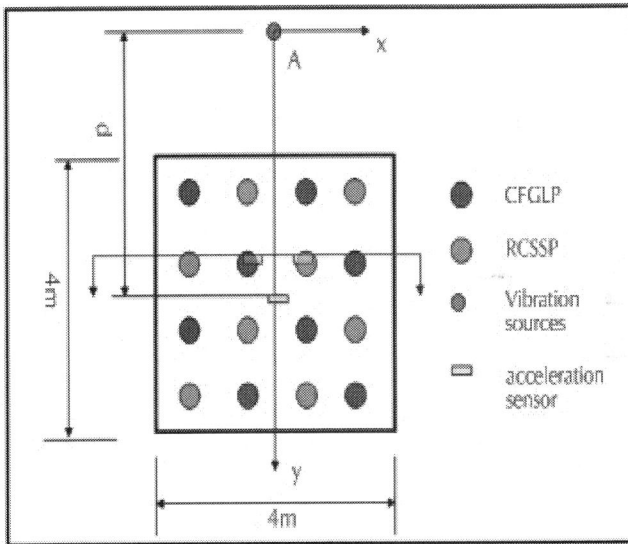

Figure 3: The arrangement of the piles and measuring elements of the Model 3. (a) Plane arrangement ;(b) 1-1profilearrangement.

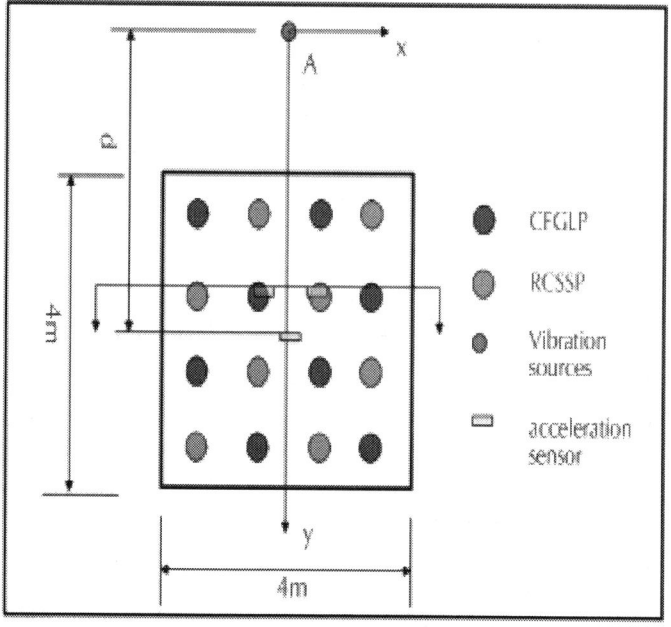

Figure 4: The p - s curves of the composite foundation.

The Main Frequency of Vibration

Field test shows that, under the same blast energy, vibration position, properties of foundation soil and form of composite foundation, have influence on the main frequency of composite foundation, but the rule is not obvious. Vibration main frequency depends on the properties of the soil and pile body between the vibration source and measuring point. When no load, the horizontal vibration main frequency of the natural foundation is at 8.2 - 9.03 Hz, and the vertical vibration main frequency is at 7.75 - 9.16 Hz; the horizontal vibration main frequency CFGP composite foundation is at 21.1 - 27.7 Hz, and the vertical vibration frequency is at 6.8 - 31.6 Hz; the horizontal vibration main frequency of the CFGLP-CFGSP composite foundation is at 17.4 - 29.6 Hz, and the vertical vibration main frequency is at 7.2 - 9.4 Hz; the horizontal vibration main frequency of the CFGLP-RCSSP composite foundation is at 7.6 - 38.4 Hz, and the vertical vibration main frequency is at 8.2 - 43.1 Hz. Horizontal vibration main frequency greater than the vertical vibration main frequency, and the vertical main vibration frequency close to the natural frequency of composite foundation. With the load increasing, the main vibration frequency of the CFGLP-RCSSP composite foundation decreases. When the load is 180 kPa, the horizontal vibration main frequency of the CFGLP-RCSSP composite foundation is at 7.4 - 22.6 Hz, and the vertical vibration main frequency is at 7.3 - 28.0 Hz.

Peak Acceleration Results

Figures 5-7 are the peak acceleration with the horizontal distance r from the measuring point to the vibration source without no pilling load on the CFGP composite foundation, when the depth of the vibration source is 6 m and the explosive quantity is 1.05 kg. From Figure 4, in addition to individual point the horizontal peak acceleration is greater than the vertical peak acceleration, and with the increase of r, the difference gradually decreases. From Figures 5-7, outside the scope of the composite foundation, the peak acceleration significantly decreased with the increase of the pilling load; in the surface of compound foundation, pilling load action makes the peak acceleration gently.

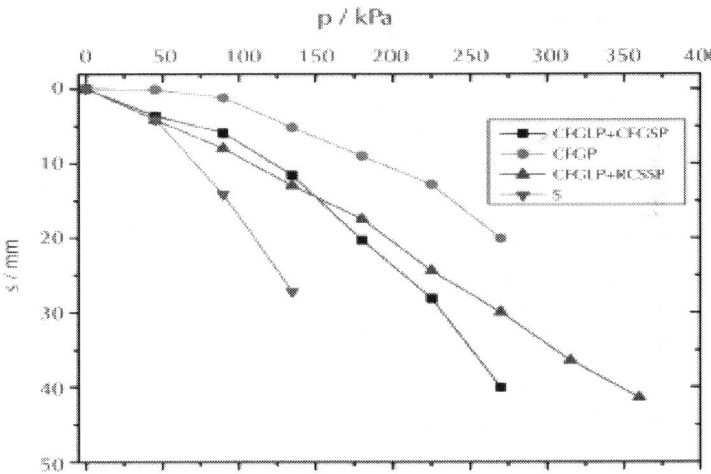

Figure 5: $a_{xmax}(a_{zmax})$-r of CFGP composite foundation (no load).

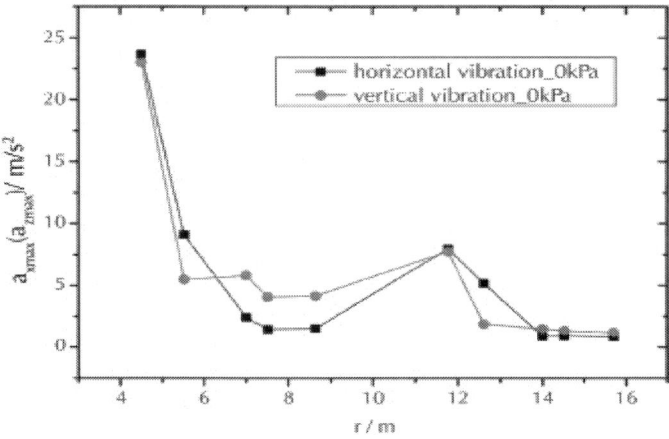

Figure 6: a_{xmax}-r of CFGP composite foundation.

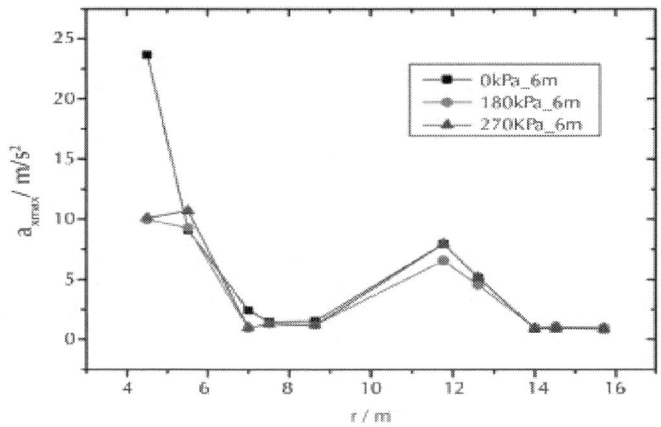

Figure 7: a_{zmax}-r of CFGP composite foundation.

When the depth of the vibration source is 6 m and r = 7 - 8.6 m, with the no-load, the horizontal acceleration peak of the CFG pile composite foundation surface is at 2.44 - 1.53, and the vertical acceleration peak is 5.8 - 4.2; while pilling load is 270 kPa, the horizontal acceleration peak is 1.00 - 1.17, and vertical acceleration peak is 4.01 - 3.96.

When the distance from vibration source to the center of CFGP composite foundation distance is 14 m and the depth of vibration source is 6 m from the ground, the ratio of horizontal acceleration and vertical acceleration peak is at 0.61 - 2.75, the measuring point outside CFGP composite foundation when the distance r is 1 m, the ratio was 1.03, and near to 1.0. When the distance from vibration source to the center of CFGP composite foundation distance is 7 m and the depth of vibration source is 6 m from the ground, the ratio of horizontal acceleration and vertical acceleration peak is at 0.36 - 1.65; the farther the horizontal distance r from the measuring point to vibration source is, the greater the ratio.

Figures 8 and 9 are the peak acceleration with the horizontal distance r from the measuring point to the vibration source without no pilling load on the CFGLP-CFGSP composite foundation, when the depth of the vibration source is 6 m and the explosive quantity is 1.05 kg. From Figures 8-10, the peak acceleration along the CFGLP on the CFGLP-CFGSP composite foundation is near to peak acceleration of the CFGSP.

When the depth of the vibration source is 6 m and r is at 7 - 8.6 m, with the no-load, the horizontal peak acceleration of the CFGLP-CFGSP composite foundation surface is at 9.05 - 1.53 m/s², and the vertical peak acceleration is 13.27 - 2.72 m/s²; while r is at 14 - 15.7 m, the horizontal peak acceleration is 5.23 - 2.25 m/s², and vertical peak acceleration is 5.13 - 0.7 m/s².

When the distance from vibration source to the center of CFGLP-CFGSP composite foundation distance is 14 m and the depth of vibration source is 6 m from the ground, the ratio of horizontal acceleration and vertical acceleration peak is at 1.02 - 3.15, measuring point distance from the vibration source is equal, the ratio is 1.49 on the CFGSP measuring point, and ratio is 3.15 on the CFGLP measuring point.

When the distance from vibration source to the center of CFGLP-CFGSP composite foundation distance is 7 m and the depth of vibration source is 6m from the ground, the ratio of horizontal acceleration and vertical acceleration peak is at 0.69 - 1.91, measuring point distance from the vibration source is equal, the ratio is 1.91 on the CFGSP measuring point, and ratio is 1.71 on the CFGLP measuring point. The farther the horizontal distance r from the measuring point to vibration source is, the greater the ratio.

Figures 11-18 are CFGLP and RCSSP peak acceleration changes of the CFGLP-RCSSP composite foundation with vibration source depth and pilling load, when the explosive quantity is 1.05 kg and pilling load is 360 kPa.

Figure 11 is the distribution law of the horizontal vibration peak acceleration on the CFGLP of the CFGLPRCSSP composite foundation. The depth of the vibration source is separately 2.5 m and 6 m, the horizontal vibration peak acceleration distribution is almost consistent. When the depth location of the vibration source is 7 m, the horizontal vibration peak acceleration maximum is at $z = 4$ m; when the location of the vibration source is 14 m, the horizontal vibration peak acceleration maximum is at $z = 0$ m.

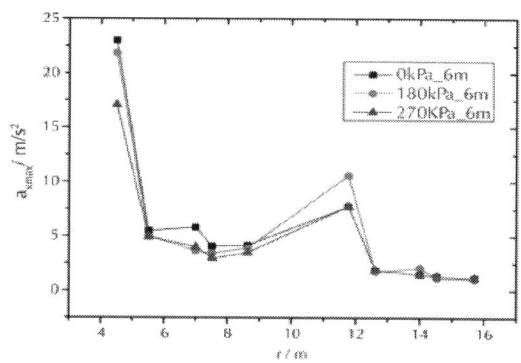

Figure 8: a_{xmax}-r of CFGLP-CFGSP composite foundation.

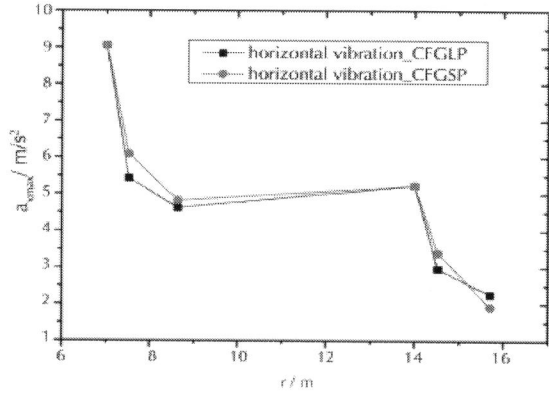

Figure 9: a_{zmax}-r of CFGLP-CFGSP composite foundation.

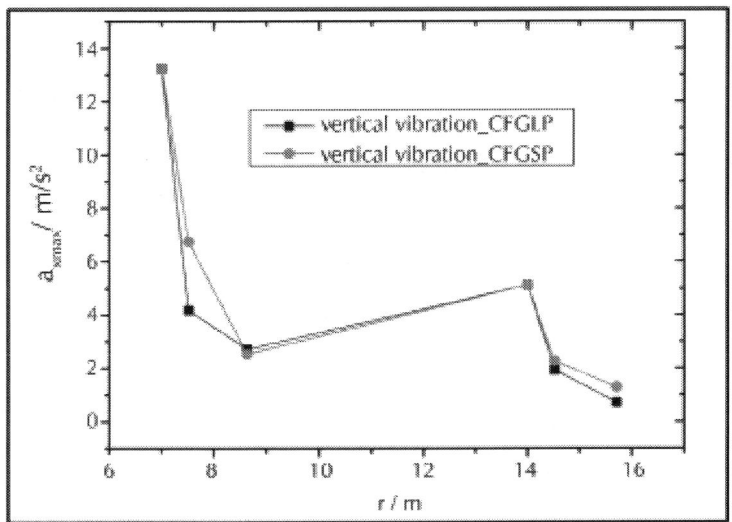

Figure 10: $a_{xmax}(a_{zmax})$-r of CFGLP-CFGSP composite foundation.

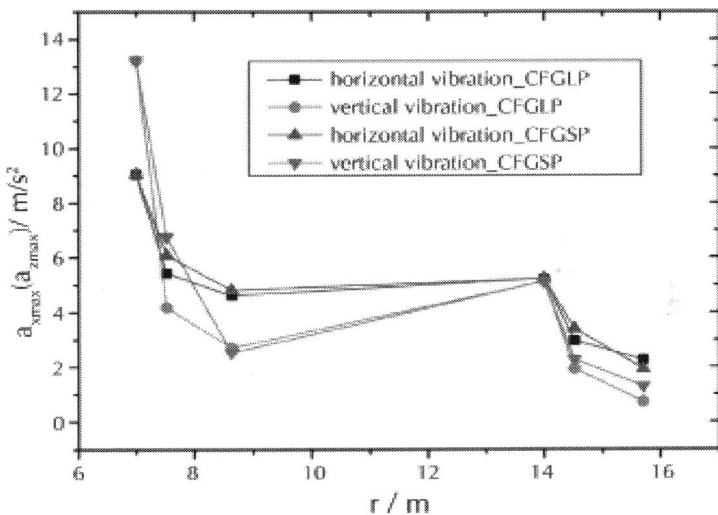

Figure 11: The peak acceleration a_{xmax} along the CFGLP.

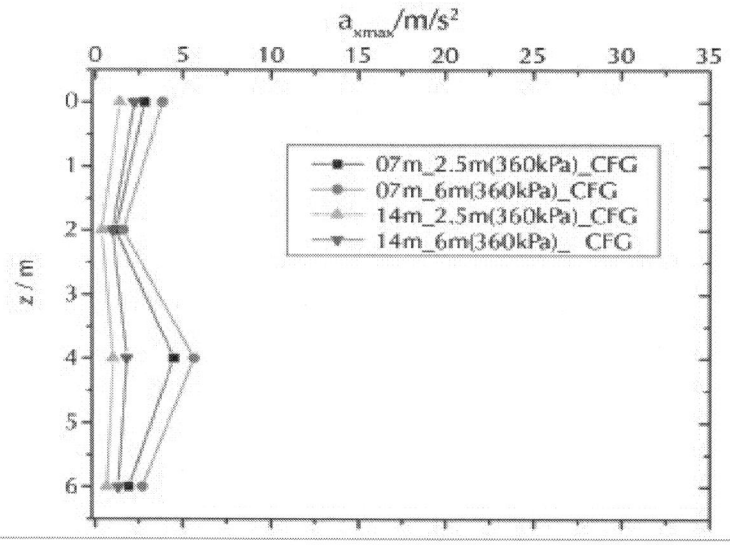

Figure 12: The peak acceleration a_{xmax} along the RCSSP.

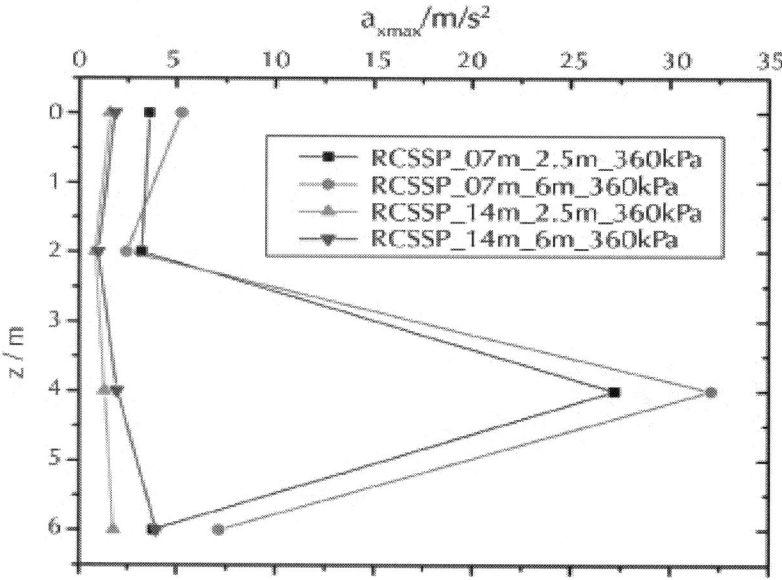

Figure 13: The peak acceleration a_{zmax} along the CFGLP.

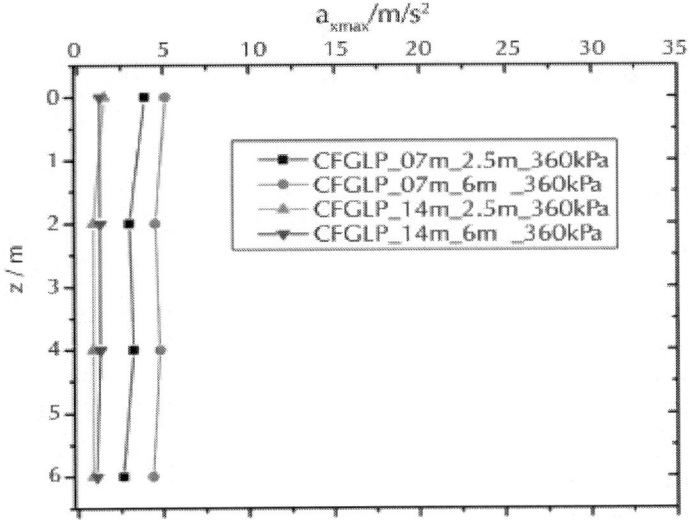

Figure 14: The peak acceleration a_{zmax} along the RCSSP.

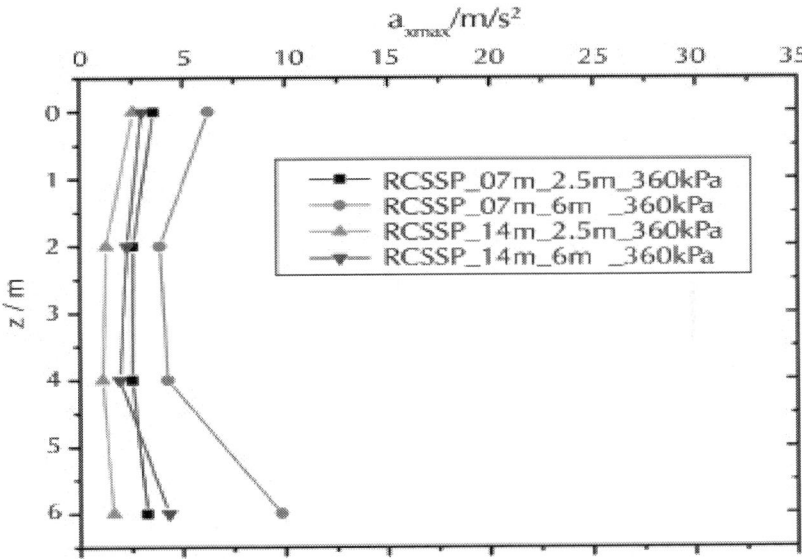

Figure 15: a_{xmax}-z in CFGLP-RCSSP composite foundation.

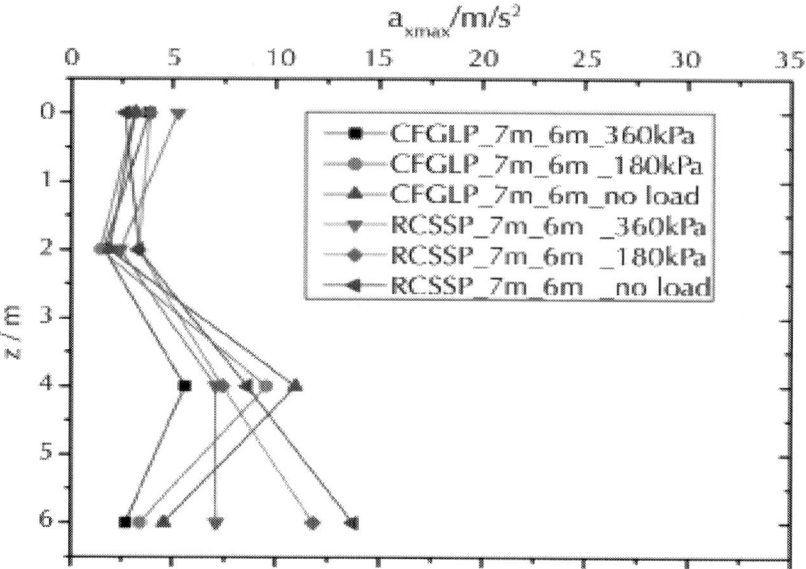

Figure 16: a_{zmax}-z in CFGLP-RCSSP composite foundation.

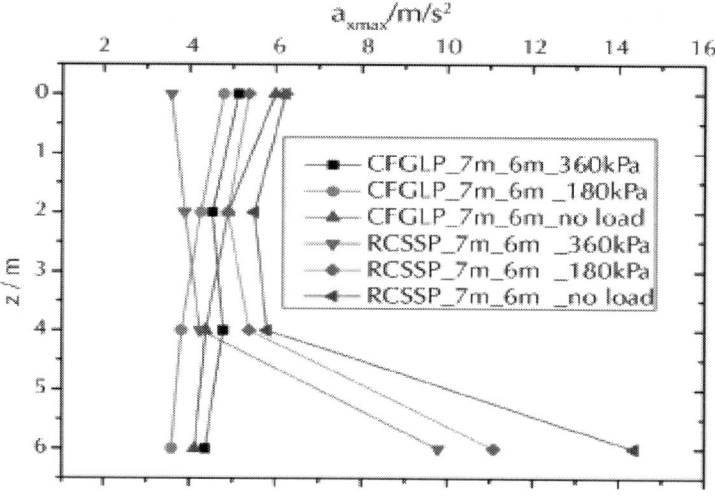

Figure 17: a_{xmax}-z in CFGLP-RCSSP composite foundation.

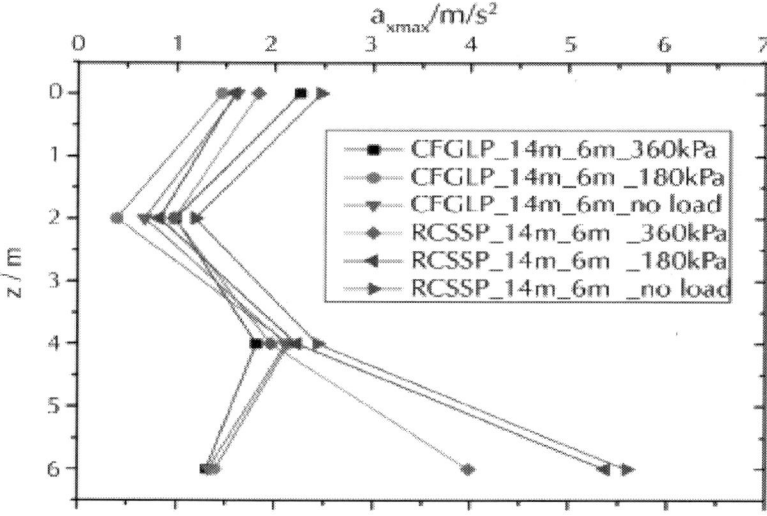

Figure 18: a_{zmax}-z in CFGLP-RCSSP composite foundation.

Figure 12 is the distribution law of the horizontal vibration peak acceleration on the RCSSP of the CFGLPRCSSP composite foundation. When the depth location of the vibration source is 7 m, the horizontal vibration peak acceleration maximum is at z = 4 m; when the location of the vibration source is 14 m, the horizontal vibration peak acceleration maximum is at z = 6 m.

Figure 13 is the distribution law of the Vertical vibration peak acceleration on the CFGLP of the CFGLPRCSSP composite foundation. The vertical vibration acceleration peak on the CFGLP is change little, and the vertical vibration peak acceleration maximum is at z = 0 m.

Figure 14 is the distribution law of the Vertical vibration peak acceleration on the RCSSP of the CFGLPRCSSP composite foundation. The vertical vibration acceleration peak on the RCSSP is change little, and the vertical vibration peak acceleration maximum is at z = 0 m.

Figures 14-18 are the distribution law of the vibration peak acceleration along the pile of the CFGLP-RCSSP composite foundation. With the increase of pilling load on CFGLP-RCSSP of the composite foundation the peak acceleration is reduced, the change range of composite foundation of 4 m pile body acceleration is consistent, the peak acceleration of the measuring point of the RCSSP is larger than that of the CFGLP. The maximum of the horizontal acceleration peak occurs in $z = 4$ m or $z = 0$ m. The maximum of the vertical acceleration peak occurs in $z = 0$ m.

CONCLUSIONS

The nonlinear degree of the p - s curves of combined pile composite foundation decreases, and CFGLP-RCSSP is closed to linear relation. The bearing capacity of the four composite piles of the CFGP, CFGLP-CFGSP, and CFGLP-RCSSP in the site are separately 225 kPa, 179 kPa, and 197 kPa, separately increases 150%, 98.8% and 119% compared to the natural Foundation.

The field test shows that, under the same blast energy, vibration source position, form of composite foundation and properties of foundation soil, influence the main frequency of the composite foundation, but the rule is not obvious. The vibration main frequency is mainly depended on properties of foundation soil and piles between vibration source and measuring point, pilling load value. Horizontal vibration main frequency greater than the vertical vibration main frequency and the vertical vibration main frequency close to the first-order natural frequency of composite foundation. With the pilling load increasing, the CFGLP-RCSSP pile composite foundation combined frequency decreased.

The field test shows that, under the same blast energy, vibration source position, the acceleration peak on the CFGP composite foundation is less than CFGLP-CFGSP the corresponding values, as the load increases, the peak acceleration gently. CFGP composite foundation is favorable on seismic.

Field test shows that, under the same blast energy, vibration source positions, form of composite foundation, properties of foundation soil and pilling load have a significant effect on the peak acceleration of composite foundation. The distribution of peak acceleration is consistent within 4 m from pile top in the CFGLP-RCSSP composite foundation. The maximum of the horizontal acceleration peak along the pile body occurs at a distance of pile top 4 m or the pile top, and that of vertical acceleration peak occurred at a pile top.

FUNDING

This work was supported by The Natural Science Foundation of Hebei Province under Grant (No. 2011- E080601).

REFERENCES

1. (2002) GB50007-2002 Code for Design of Building Foundation(s). China Architecture Industry Press, Beijing.

2. Yoshio, S. (1983) Deep Mixing Chemical Method Using Cement as Hardening Agent. Symposium on Soil and Rock Improvement Techniques, Bangkok, 34-37.

3. Horii, N., Toyosawa, Y., Tamate, S. and Hashizume, H. (1998) Stability of Composite Ground Improved by Deep Mixing Method. Proceedings of 2nd International Conference on Ground Improvement Techniques, Singapore, 193- 198.

4. Zheng, G. and Jiang, X.L. (1999) Research on the Bearing Capacity of Cement Treated Composite Foundation. Rock and Soil Mechanics, 3, 46-50.

5. Ma, H.L. (2003) Quantitative Analyses of the Influence of Pile Length and Other Factors on Capability and Modul of Cement Stabilized Soil Composite Foundation. Chinese Journal of Geotechnical Engineering, 11, 720-723.

6. Fu, J.H. and Song, E.X. (2000) Analysis of Rigid Pile Composite Foundation's Working Performance. Rock and Soil Mechanics, 21, 335-339.

7. Wang, M.-S., Wang, G.-C., Yan, X.-F., et al. (2005) In-Situ Tests on Bearing Behavior of Multi-Type-Pile Composite Subgrade. Chinese Journal of Geotechnical Engineering, 27, 1142-1145.

8. Yan, M.L., Wang, M.S., Yan, X.F. and Zhang, D.G. (2003) Study on the Calculation Method of Multi-Type-Pile Composite Foundation. Chinese Journal of Geotechnical Engineering, 25, 352-355.

9. Ding, J.H. (2007) Reliability Analysis on the Bearing Capacity of Composite Foundation with Multi-Type Compound Piles. Engineering Mechanics, 25, 168-172.

10. Ding, J.H., Liu, F.R. and Du, E.X. (2008) Dynamic Characteristic Analysis on Composite Foundation with Soil-Cement Piles and CFG Piles. Fly-Ash Comprehensive Utilization, 6, 37-40.

11. Wang, W.Y., Zhao, T. and Meng, Y.J. (2012) The Numerical Analysis on Dynamic Characteristics of CFG Pile Composite Foundation under Blasting. Engineering Mechanics, 29, 150-155.

12. Wang, W.Y., Zhao, T. and Ding, J.H. (2011) Effects on Dynamic Characteristics and Response of Rammed Soil-Cement Pile Composite Foundation. Engineering Mechanics, 28, 187-191.

13. Wang, W.Y., Zhao, T. and Ding, J.H. (2010) Influence Factors of Dynamic Characteristics and Response of CFG Pile Composite Foundation. Chinese Journal of Geotechnical Engineering, S2, 115-118.

14. Zhao, T., Yang, C.M. and Wang, W.Y. (2010) The Dynamic Test Research on CFG Pile Composite Foundation under Blasting. Journal of Highway and Transportation Research and Development (Applied Technique), 7, 121-122.

15. Ding, J.H., Wang, W.Y., Zhao, T., et al. (2013) the Dynamic Characteristic Experimental Method on the Composite Foundation with Rigid-Flexible Compound Piles. Open Journal of Civil Engineering, 3, 94-98. http://dx.doi.org/10.4236/ojce.2013.32010

Key Techniques and Risk Management for the Application of the Pile-Beam-Arch (PBA) Excavation Method: A Case Study of the Zhongjie Subway Station

Yong-ping Guan,[1] Wen Zhao,[1] Shen-gang Li,[1] and Guo-bin Zhang[2]

[1]School of Resources and Civil Engineering, Northeastern University, Shenyang 110819, China

[2]Shenyang Design and Research Institute of Municipal Engineering, Shenyang 110015, China

ABSTRACT

The design and construction of shallow-buried tunnels in densely populated urban areas involve many challenges. The ground movements induced by tunneling effects pose potential risks to infrastructure such as surface buildings, pipelines, and roads. In this paper, a case study of the Zhongjie subway station located in Shenyang, China, is examined to investigate the key construction techniques and the influence of the Pile-Beam-Arch (PBA) excavation method on the surrounding environment. This case study discusses the primary risk factors affecting the environmental safety and summarizes the corresponding risk mitigation measures and key techniques for subway station construction using the PBA excavation method in a densely populated urban area.

INTRODUCTION

Subway construction projects are generally located in complex surrounding environments and are often built beside or under residentially, commercially, or officially important buildings. However, tunneling in dense urban areas can cause ground movements and surface settlements, which may lead to additional deformations and damage to existing structures and utilities such as residential buildings and pipelines [1]. In this situation, one of the key factors in design and construction problems may be the amount of allowable settlement.

Underground projects are extremely complex and are associated with many uncertainties resulting from geological and geomechanical parameters, external load, and construction quality [3]. These uncertainties during tunneling can lead to potential risks to both the workers and the surrounding environment [4]. To minimize the adverse effects on the surrounding environment and perform appropriate risk mitigation measures in time, a risk management technique should be adopted throughout the underground construction project development [5].

Many studies have been conducted to investigate the risk assessment of tunneling projects in urban areas and the adverse impact of the PBA excavation method on ground settlement and adjacent utilities.

Considering the uncertainties, Einstein [6] proposed a probabilistic approach to assist decision making in the form of engineering design, selecting particular construction procedures or more general decisions made by decision makers in geotechnical engineering. Reilly [7] discussed an overview of management for complex, underground, and tunneling projects, suggested an improved methodology for the "project delivery process," and summarized crucial supporting systems such as partnering and risk mitigation. You et al. [8] presented a methodology to select an optimal supporting scheme and advance rate quantitatively for the design of a tunnel by performing a risk analysis considering the construction fee and the cost of losses related to tunnel collapse. Fang et al. [9] proposed a risk management methodology that aims at process control for ground settlement and surface buildings, to guarantee the environmental safety.

In terms of the PBA excavation method research, Wang et al. [10–12] analyzed the influence of a subway station constructed by the PBA excavation method on the ground settlements and adjacent pipelines by means of numerical simulation and field measurements. Yang et al. [13] optimized the PBA construction procedure by implementing different schemes for heading opening patterns and heading excavation sequences based on three-dimensional numerical modelling. He [14] studied the main theoretical problems encountered during construction and the influence of metro tunneling using the PBA method on adjacent piles using the numerical simulation method.

However, there is limited research about the environmental risk assessment for subway station construction using the PBA (Pile-Beam-Arch) excavation method. This paper presents an in-depth investigation of the influence of the PBA excavation method on ground movements and the surrounding environment. Simultaneously, the potential risk encountered in this project during construction and corresponding risk mitigation measures are elaborated in detail. It is helpful to provide a valuable experience for other shallow-buried subway station construction projects in densely populated urban areas.

PROJECT OVERVIEW

Geographic Location

The first subway project constructed in Shenyang was the Metro Line 1 Project-Blue Line, China (see Figure1). Metro Line 1 is almost 22 km in length with 22 subway stations. The line was located along jammed roads in the central city area as shown in Figure 1. The line runs from the Thirteen Street Station in the west to Li-Ming Square Station in the east. The strata in Shenyang along the tunnel alignment is comprised of numerous sandy layers with variable grain size distribution from silt sand to course gravel.

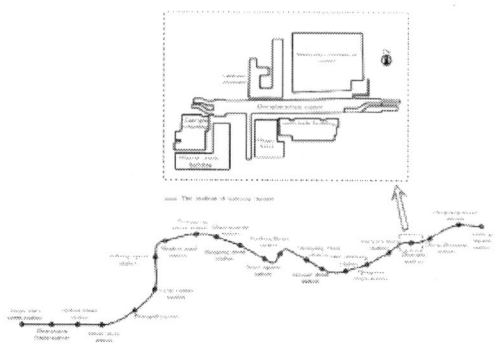

Figure 1: Main route of Shenyang Metro Line 1.

In this study, an environmental risk assessment was performed for the Zhongjie subway station, which was constructed using the PBA excavation method and is located in the chainage of DK17+880.567 to DK18+114.267. Zhongjie station is a double-deck tunnel of double-arch shape and the cross section size is approximately 19.7 m × 15.85 m. The overburden thickness is approximately 8.59 m, and the buried depth of the bottom is approximately 24.89 m. A typical cross section of the Zhongjie station is shown in Figure

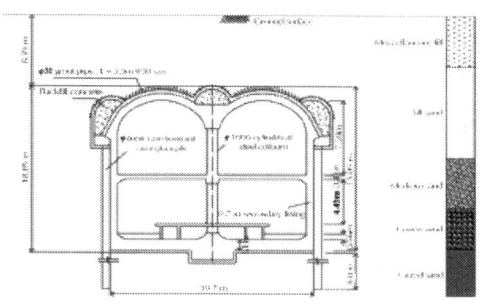

Figure 2: Typical cross section of Zhongjie subway station.

PBA Excavation Method

Urban subway stations are generally shallowly buried in China. Because of a thin overburden layer, a "soil arching effect" cannot be adequately formed above the tunnel roof [15]. Furthermore, the collapse surface of a shallow tunnel will easily extend from the tunnel face to the ground surface. However, the adjacent utilities (pipelines, crowded surface buildings) are significantly sensitive to ground movements induced by tunnel excavation, and excessive ground movements may induce damage to adjacent utilities. Plenty of successful subway construction projects in China have proved that the shallow tunneling method (STM) is very suitable for shallowly buried tunnel constructed in urban areas, which is inconvenient to excavate using the cut and cover method.

The PBA (Pile-Beam-Arch) excavation method is one major approach to shallow tunneling and was adopted in the Zhongjie subway station construction project. Zhongjie subway station is a double-arch-double-span-double-deck station. The main steps of the PBA excavation method for the Zhongjie subway station are shown in Figure 3. The excavation sequence of the typical cross section can be divided into five steps. It is necessary to note that the PBA excavation process was implemented under dry conditions. Thus, an appropriate dewatering scheme should be designed to ensure that the groundwater table remains at least 0.5 m below the excavation bottom of the station during construction in similar situations.

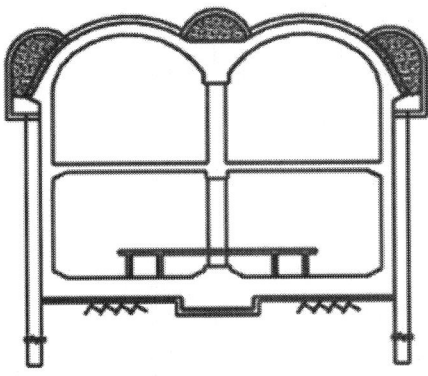

Figure 3: PBA excavation method adopted in the Zhongjie subway station.

The main construction sequences of the PBA excavation method are as follows:(1)upper pilot heading excavation using a multiface tunneling method (see Figure 3(a));(2)concreting the side piles (see Figure 3(b));(3)excavating the lower pilot heading (see Figure 3(c));(4) installing the middle steel column and constructing supporting arch (see Figure 3(d));(5)excavating the main structure via a top-down construction method and concreting secondary lining (see Figure 3(e)). The construction site situation is shown in Figure 4.

Figure 4: Photo of construction site. (a) First floor and (b) Second floor.

NUMERICAL INVESTIGATION OF THE PBA METHOD

Three-Dimensional Modelling of Tunnel

The PBA excavation procedure was investigated with a three-dimensional simulation implemented using Flac3D code according to the finite difference method. The numerical simulation was capable of investigating the influence of PBA excavation processes on ground movements, which is helpful for providing a basis for the establishment of a process control index for ground surface settlements. In this study, the tunnel was assumed to be excavated in green field condition (ground surface above tunnel with no existence of buildings), and auxiliary methods such as sleeve-pipe grouting, forepoling pipe grouting, and feet-lock bolt were not considered. In addition, the following hypotheses were adopted in this study:(1)soil properties are assumed to be homogeneous and isotropic, with an elastic perfectly plastic constitutive relation with a Mohr-Coulomb yield criterion;(2) supporting structures such as lining and steel column are considered as elastic media.

The numerical simulation model was 50 m in the -direction, 120 m (approximately 6D, D is the tunnel span) in the -direction, and 60 m in the -direction, and the number of grid cells was 741040 (as shown in Figure6). The model was meshed in 50 longitudinal blocks of the same size. The tunnel linings were assumed to remain in contact with the surrounding soils and installed immediately after excavation. The hardening process of the sprayed concrete and the time delay effect of primary lining construction were considered by reducing the elastic modulus of the primary lining. In this study, the elastic modulus of the primary lining is 1/10 of that of its real value according to the suggestion given by Fang et al. [9].

As for boundary conditions, the horizontal displacements were set to zero at each side, which means that vertical displacements were allowed, and the node at the bottom of the mesh was fixed in both the vertical and horizontal directions. The top surface of the model was free in both directions.

In this project, the side pile's diameter is 0.8 m with a clearance of 0.4 m. By considering the importance of the bending rigidity of side pile, the equivalent values of EI were modelled in the rectangular cross section (see Figure 5). Thus, the equivalent thickness of the rectangular wall can be derived from (1) as follows:

$$\frac{1}{12}(D+t)h^3 = \frac{1}{64}\pi D^4 \tag{1}$$

Where D is the pile's diameter, h is the pile spacing, which is 0.4 m in this project, and is the equivalent thickness of the rectangular wall.

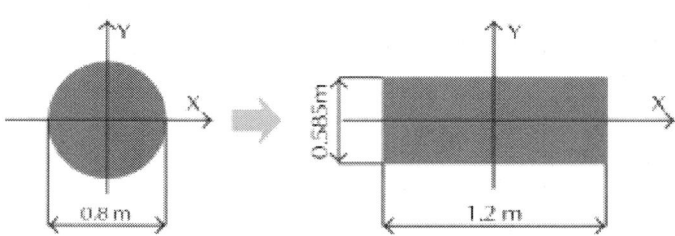

Figure 5: Modelling piles as a continuous wall.

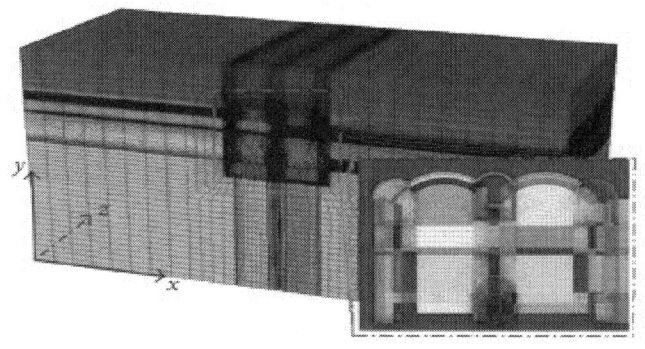

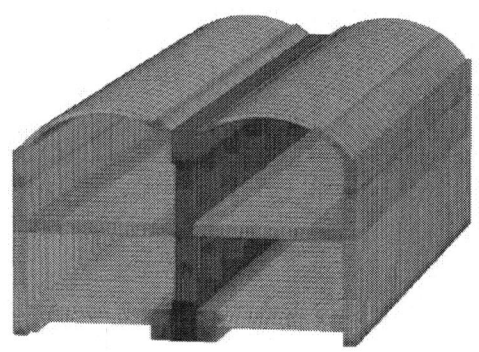

Figure 6: Finite element model. (a) 3D model and (b) Main structure.

The soil properties and related parameters of the supporting structures used in this study are summarized in Table 1.

Table 1: Mechanical parameter of soil layer and supporting structure

Name	h(m)	E(Mpa)	c(kPa)	φ (°)	γ (kN/m)	μ
Miscellaneous fill	6.20	7.94	5	13	20.1	0.45
Silt sand	8.8	19.8	8.8	30.7	17.36	0.28
Medium sand	6.80	27.4	21.5	30.7	18.15	0.26
Coarse sand	4.4	27.4	21.5	34.7	18.15	0.26
Gravel sand	5.9	98	1.7	36.7	20.0	0.23
Primary lining	0.3	3000	/	/	25	0.25

Results of the Numerical Simulation

For the subway station construction projects, especially in dense urban areas, it is necessary to prevent surface existing buildings and underground utilities from failing due to the ground movements induced by tunneling. The ground settlements should be strictly controlled, as excessive ground settlements may cause tunnel cave-in and cause negative effects or even damage to the existing surface buildings. Therefore, it is of paramount importance to investigate the influence area and degree as a response of a subway station construction using the PBA excavation method.

The section at m was chosen for investigation in this study. The strata settlement trough and ground movements are presented in Figures 7 and 8. Figure 7 shows the simulated vertical displacement contour after tunnel excavation. Figure 7(a) indicates that the width of the excavation influence region was found to be 3.0 times the tunnel span, approximately 60 m. This means the surface buildings located within a 30 m distance from the tunnel centerline would inevitably be affected by the tunneling activities and subject to differential settlements.

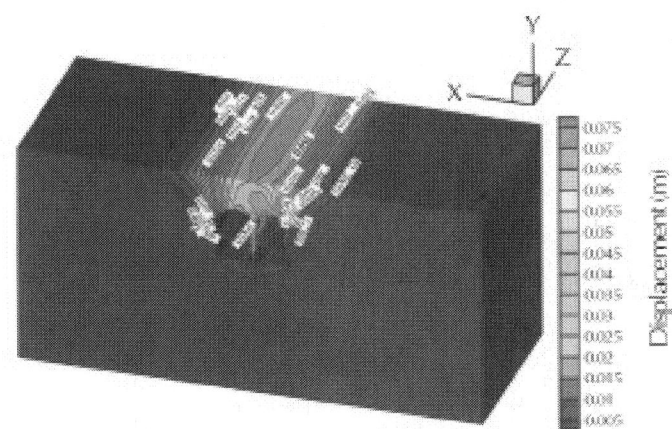

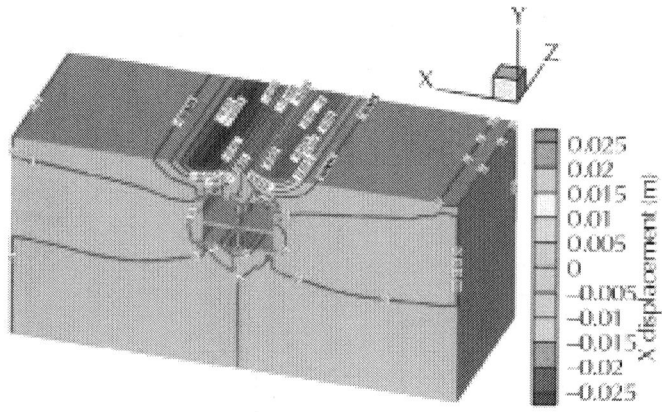

Figure 7: Displacement field obtained from the numerical simulation. (a) Contour of ground settlement after excavation and (b) Horizontal displacement of stratum after excavation.

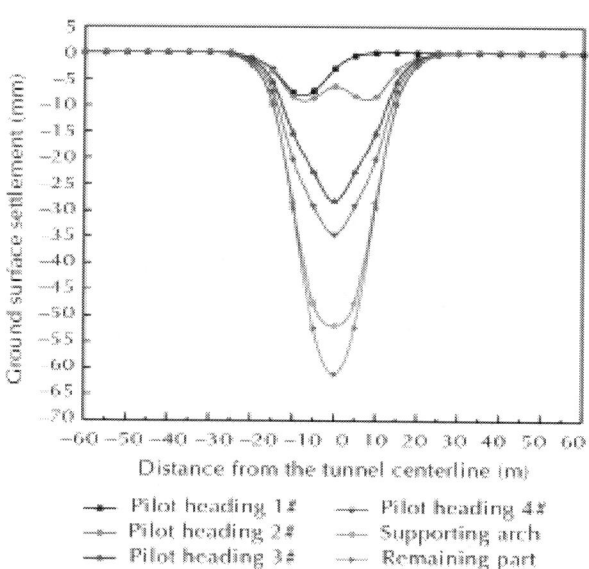

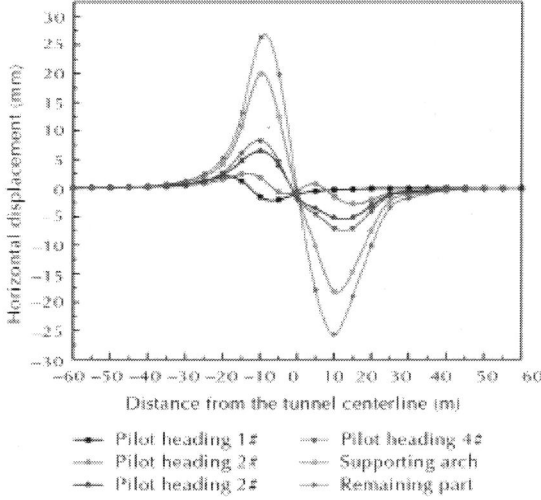

Distance from the tunnel centerline (m)

- Pilot heading 1#
- Pilot heading 2#
- Pilot heading 2#
- Pilot heading 4#
- Supporting arch
- Remaining part

Figure 8: Ground movements induced by tunnel excavation in section m. (a) Ground surface settlements and (b) Horizontal displacement of ground surface

Figure 7(b) shows that the horizontal displacement of the ground surface was relatively symmetric about the tunnel centerline.

The influence of the PBA excavation method on the ground movements is shown in Figure 8, in which the progressive development of the transverse surface settlement trough and horizontal displacement of ground surface for the monitoring section $z = -30$ m are shown. The figure shows that the excavation influence zone extends symmetrically for approximately 30 m from the tunnel axis. As seen from Figure 8(a), the maximum ground settlement was approximately 61.1 mm after the excavation, appearing above the tunnel centerline. As shown in Figure 8(a), the additional ground surface settlement caused by the 1#, 2#, 3#, and 4# pilot heading excavation was 2.99 mm, 3.43 mm, 21.78 mm, and 6.4 mm, respectively. The maximum ground surface settlement was reached 34.6 mm after the excavation of pilot headings. In addition, the excavation of the supporting arch also has a great influence on the ground settlements, which caused 17.3 mm of additional settlement. The excavation of the four pilot headings and supporting arch had a major influence on the ground surface settlements, approximately

56.6% and 28.3% of additional settlement, respectively. This trend is similar to those reported by Wang et al. [11]. Such a trend indicates a requirement for closer attention to the stability of surface buildings when excavating pilot headings and supporting arches, especially during the period immediately after the installation of the sprayed concrete, which has not yet achieved its final stiffness.

As presented by Liu et al. [16], the horizontal displacement of ground surface induced by tunneling projects can also lead to the differential settlement and cracking of existing building. The horizontal displacement of ground surfaces caused by the PBA excavation method is shown in Figure 8(b), which is distributed in a wave shape. The horizontal displacements of ground surface were perfectly symmetrical about the tunnel centerline during the construction process. The maximum horizontal displacement was approximately 25 mm and appeared at 10 m distance from the tunnel centerline. The horizontal displacement of the ground surface induced by the pilot heading and supporting arch excavation constitutes a high proportion of the total displacement.

POTENTIAL RISKS IDENTIFICATION

Groundwater Drawdown

The construction of an underground structure below the water table requires a strict and elaborate dewatering scheme. Groundwater may undermine construction safety since it induces additional loads on tunnel linings and decreases the soil strength [4]. The collapse of the Seoul subway tunnel [17] and that of a subway station in Shanghai have demonstrated the significant adverse effect of groundwater on underground projects. The presence of groundwater has increased the potential risks of underground projects.

There are rich groundwater resources in the Shenyang area, and the groundwater level is approximately 8 m below the ground surface. The Zhongjie station is located in sandy layers, which are strongly permeable, and the groundwater recharge velocity is fast. Thus, groundwater control is a key element in this project. The conventional handling of groundwater control may be performed using several methods: grouting, pumping, diaphragm wall, or a combination of them [18]. For subway construction projects, pumping wells have been extensively used in the Shenyang area because of the flexibility and low cost. However, the vast majority of pumping groundwater activities may lead to groundwater leakage and ground settlement due to the increase of soil effective stress. The design of the dewatering system should consider the environmental risks induced by groundwater drawdown.

Super-Shallow-Buried Depth

A general classification of tunnel cover-to-span is proposed by Wang [19] for the purpose of evaluation of the tunneling effect on ground settlement, based on the C/S ratio (cover-to-span). It is considered to be a shallow-buried tunnel when the C/S (cover-to-span) is within a range of 0.6–1.5, whereas, when the C/S is smaller than 0.6, it is considered to be a super-shallow-buried tunnel. For a deep-buried tunnel, there is a "soil arching effect" formed over the roof of the tunnel, which supports a large portion of overburden loads [20]. However, the mechanical behavior and deformation of the shallow-buried tunnel constructed in soft ground is significantly different from deep-buried tunnels, as the overburden layer is too thin to form a "soil arching effect" in shallow-buried tunnel [21].

In this project, the overburden layer is approximately 8.59 m, and the tunnel span is 19.7 m, for a C/S ratio of 0.43, a level at which it is difficult to produce a "soil arching effect." Under this condition, the ground movements or collapse zones induced by excavation easily extend to the ground surface [22]. This unfavorable factor necessitated higher requirements for the tunnel excavation method, support patterns, water drainage, and grouting activities, and also increased the construction difficulty.

Existing Surface Buildings

Zhongjie subway station was constructed underneath a densely populated urban area, where many sensitive surface buildings might be affected by even minor variations in the foundation conditions, caused by either ground movements or dewatering activities. The location of surface buildings relative to the Zhongjie station is illustrated in Figure 9, which also shows the location of the settlement monitoring points for these surface buildings. The investigated detailed information about these surface buildings is summarized in Table 2.

Table 2: Detailed conditions of surface buildings

Buildings	Foundation types/ buried depth	Number of stories	Height/m	Construction time
Guanglu Cinema	Box foundation (−6 m)	6	20	1999
Women's world building	Box foundation (−7.5 m)	4	20	1995
Meigui Hotel	Box foundation (−7.5 m)	20	70	1987
Laobian restaurant	Pile foundation (−8 m)	7	25	1992
North Trade Building	Precast pile foundation (−8.5 m)	6	30	1994
Shenyang commercial center	Strip foundation (−11.6 m)	6	30	1991

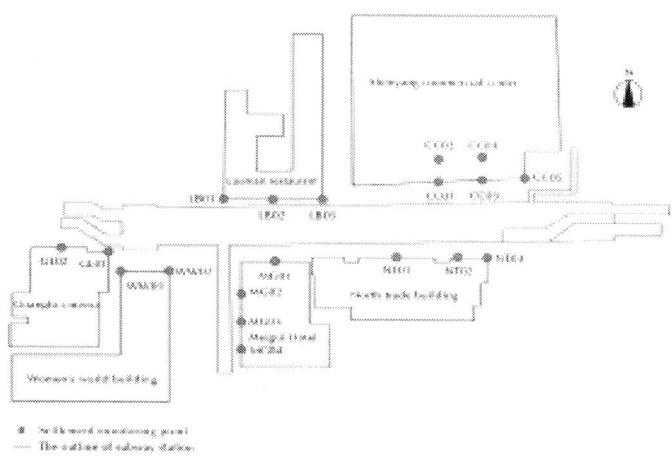

Figure 9: Zhongjie station and surface buildings relative locations.

These surface buildings were subjected to a detailed building condition investigation prior to the tunnel excavation, such as foundation type, age, height, and relative location to the subway station. According to the survey, these buildings are on different foundation types, that is, pile foundation, box foundation, and strip foundation, which were constructed in the last 15 years of the 20th century (see Table 2). Among all the cases, the Meigui Hotel is the oldest building and was constructed in 1987. The height of the building reaches 70 m, and it is adjacent to the temporary construction passage (#3 passageway) of the Zhongjie station. The distance between the exterior wall edge and excavation exterior edge is 2.67 m. The north side of the Meigui hotel is only 4.0 m away from the excavation exterior edge. Therefore, the Meigui Hotel is considered to be of the highest risk among surface buildings.

ENVIRONMENTAL RISKS MITIGATION MEASURES

The utilization of isolation piles for ground movement control has been proved effective [23]. However, its disturbance and high construction cost have limited its application in practical engineering. For the Zhongjie station project, the implementation of isolation pile techniques is very difficult because of the limited space on the ground surface. Therefore, other risk mitigation measures, such as the sleeve-valve-pipe grouting technique, double-layer pipe grouting, and forepoling pipe grouting, are adopted to ensure the safety of tunnel construction and surrounding environment. All of these methods are implemented to minimize the ground movements induced by tunneling activities. More detailed information about these measures is elaborated as follows.

Groundwater Control

Dry tunneling conditions are one of the most important preconditions for PBA tunnel construction. The dewatering scheme should be designed to guarantee that the groundwater table remains at least 0.5 m below the excavation bottom during the excavation. In addition, the settlements of the ground surface and surface buildings caused by dewatering process should be controlled within an allowable range to mitigate the environmental risk during dewatering.

In conventional dewatering approaches, the pumping well is always arranged using large spacing, a high pumping rate, and deep depth in the Shenyang area. As shown in Figure 10(a), the application of this dewatering scheme in subway station construction in densely populated urban areas has encountered some challenges. (1)To ensure the water table that between adjacent pumping wells remains at least 0.5 m below the excavation bottom during the excavation, the depths of the pumping well needed to be designed to be deep enough. This means a massive quantity of groundwater needs to be pumped, which leads to ground settlement and a substantial waste of expensive electricity.(2)Compared with normal building foundations, the subway station was composed of different building units, which had a different buried depth and excavation sequence. Thus, the application of a traditional dewatering scheme will pose potential difficulties for final-period management.(3)The subway station construction project has a characteristic of long construction period. Therefore, the long-term dewatering process requires high reliability and performance for these pumping wells. The groundwater level between adjacent pumping wells increases immediately once any one of these pumps breaks down because every single pumping well plays a good role in groundwater control.

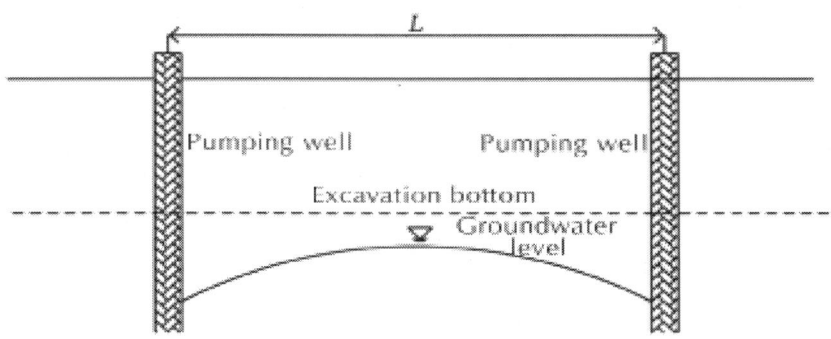

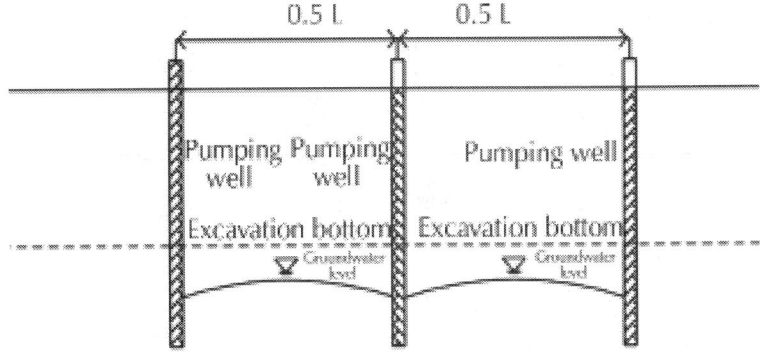

Figure 10: Details of pumping well layout. (a) Large pump with large spacing and (b) Small pump with small spacing .

Considering the above-mentioned shortcomings that exist in traditional dewatering scheme, in this project, the pumping well was arranged with small spacing (see Figure 10(b)). Compared to the above-mentioned dewatering scheme, it has the following advantages:(1)Under the same dewatering depth, the total amount of pumping water was decreased significantly, and the dewatering cost and environmental risks were also minimized.(2)Because of the small spacing, pumping wells can be flexibly arranged for different building units and construction sections. It is convenient to cease the dewatering activities for the sections where construction has been completed, and this is also helpful for decreasing the cost of dewatering.(3)Because more pumping wells are arranged around the station, the failure of a single pumping well has a limited impact on the whole dewatering system, with more time being allocated to maintaining the pumping well.

Pilot Heading Excavation Using A Multifaced Tunneling Method

As mentioned in Section 3.2, the excavation of pilot headings had a major influence on the ground surface settlements. To restrict the ground movements induced by pilot heading excavation, some risk mitigation measures were implemented during pilot heading excavation. The forepoling pipe was installed prior to the excavation to improve the stability of the excavation face and the soil properties in the front of the tunnel face. The related parameters of the forepoling pipe used in this project are presented in Figure 11: φ 32 mm with a thickness of 3.25 mm and a length of 1.8 m. These forepoling pipes were inserted at an angle of 10°–30° with the tunnel longitudinal direction into the soil above the arch ahead of the excavation face. The pilot heading was excavated using a multifaced tunneling method with short advance length. After the excavation of the upper bench, the overburden pressure exerts on the steel rib, which may lead to an integral sinking without a feet-lock bolt. Hence, the feet-lock bolts were installed immediately after excavating the upper bench to restrict the settlement and horizontal convergence of the steel rib. Luo and Chen [2] studied the complete set of the observation data collected from the tunnel construction project, and he reported that the feet-lock bolt was significantly effective in restricting the ground movement and grid steel rib's deformation in soft ground. He also illustrated the mechanism of the feet-lock bolt, as shown in Figure 12. The feet-lock bolt was subject to external forces such as bending moment, shear force, and axial force, which were transferred from the steel rib.

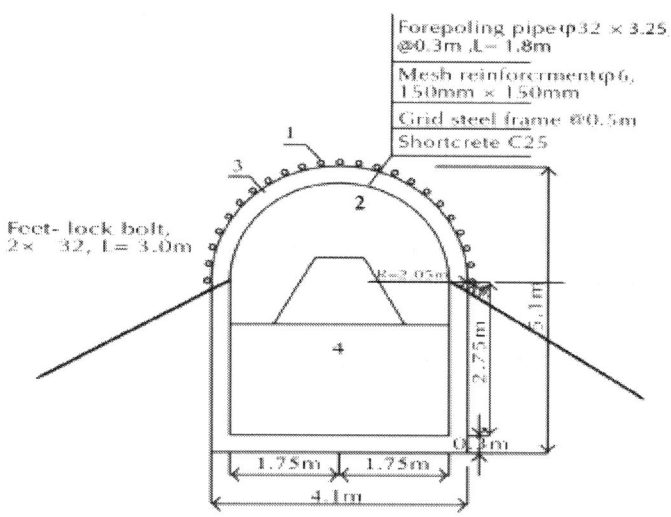

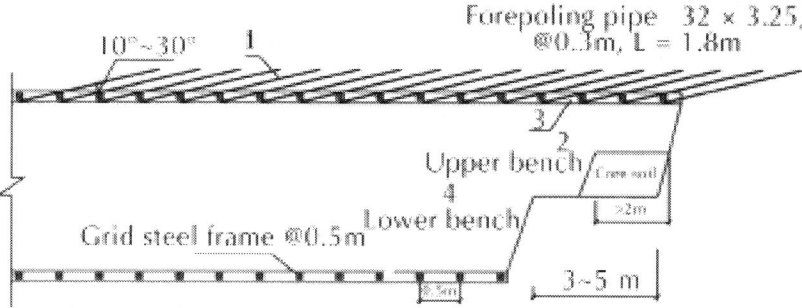

Figure 11: Excavation sequence of the pilot heading. (a) Cross section and (b) Longitudinal section.

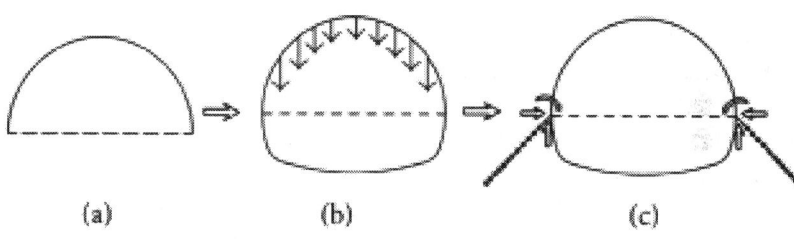

(a)   (b)   (c)

Figure 12: Mechanism of the feet-lock bolt (after [2]).

Sleeve-Valve-Pipe Grouting Technique

The sleeve-valve-pipe grouting technique has been extensively used in urban subway station construction projects in China. By injecting cement grout into the sandy strata, soil particles and grout are bonded together, and the void space within the soil particles are filled with grout [24].

The sleeve-valve-pipe grouting technique was adopted in the Zhongjie station project prior to the excavation to minimize ground settlement. The grouting was injected into the region between 3.0 m below the ground surface and 1.5 m above the tunnel crown and the thickness varies from 3.0 m to 5.0 m (see Figure 13) for the region between the subway station and existing surface buildings. Six-meter-long 70 mm grout pipes were drilled from the ground surface toward the tunnel crown. The grout injection was conducted at a rate of 50 L/min. To prevent ground uplifting, the grouting pressure was strictly controlled within a range from 0.4 Mpa to 0.6 Mpa during the construction.

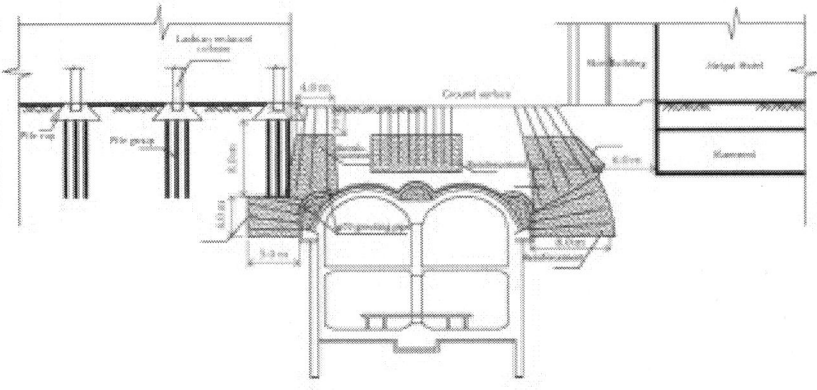

Figure 13: Sleeve-valve-pipe grouting technique.

The related parameters of grouting included grout formula $W:C=0.8:1,$, incorporation of some accelerator, gel time of approximately 40 s, and a diffusion radius of approximately 0.7 m.

During construction, there were more than 1700 grout holes placed around the tunnel, with a minimum theoretical volume of 4500 m³. To determine the grouting quality, ground penetrating radar was applied in this project. Two cross sections of the detected ground compactness are shown in Figure 14. We can observe from the monitoring graph that the soil compactness increased significantly compared with the initial conditions. This indicates that the elastic modulus and strength of the strata increased effectively and improved self-bearing ability, which is helpful for restricting ground movements.

Non-grouting (KD18 + 057)

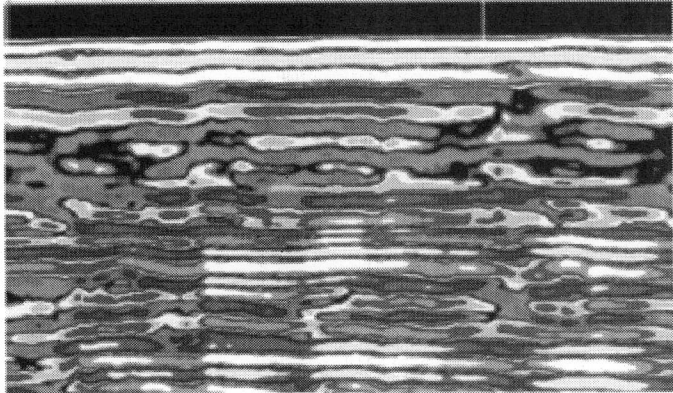

Non-grouting (KD18 + 003)

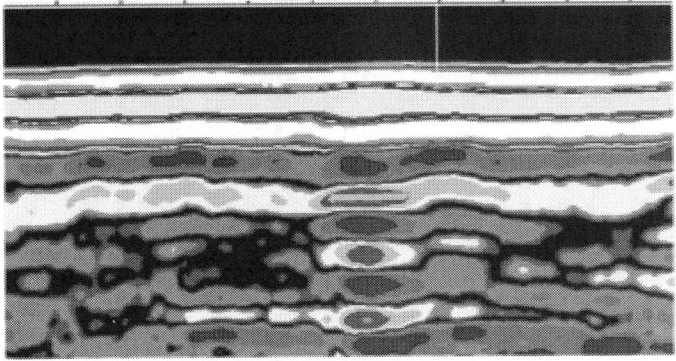

After grouting (KD18 + 057)

After grouting (KD18 + 003)

Figure 14: The comparison between grouting and non-grouting strata.

Reinforcement of the Building Foundation

The excavation of subway stations will induce the movement of the foundation soil beneath the surface building toward the tunnel, which leads to the differential settlements of buildings or even collapse. The interaction between tunnel excavation and existing structure not only poses a potential risk to the underground project but also threatens the safety of surface buildings. To ensure the safety of tunnel and surface buildings, $\varphi 70$ mm grout pipes were drilled from the inside of pilot headings toward the building foundation. The grouting pressure was controlled within 0.8 MPa to 1.0 MPa. Detailed foundation of building reinforcement schemes are illustrated in Figure 15.

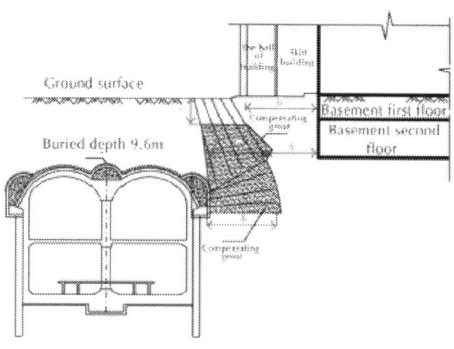

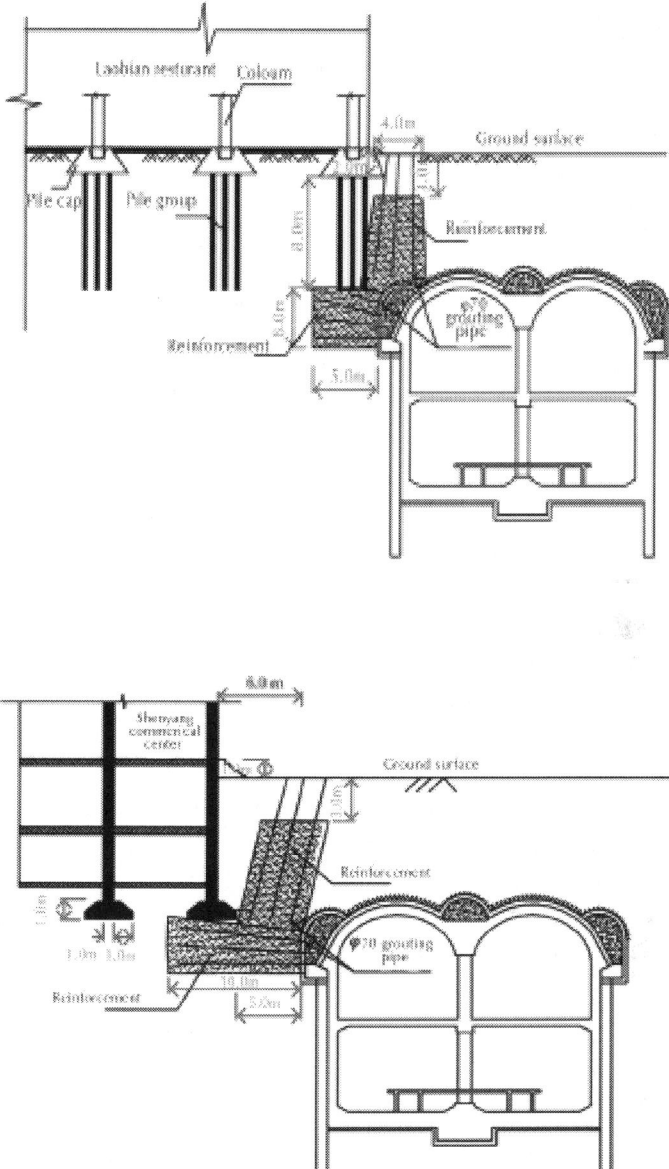

Figure 15: Reinforcement of the building foundation. (a) Meigui Hotel, (b) Laobian restaurant, and (c) Shenyang commercial center.

Double-Layer Grouting Advanced Support

The numerical simulation performed in Section 3.2 indicates that supporting arch construction is one of the key steps in the PBA excavation method. Double-layer forepoling pipes were implemented to strengthen the soil above the supporting arch prior to the excavation (see Figure 16) to ensure the safety of the tunnel and surface buildings during the supporting arch construction. The lengths of the inner layer pipe and outer layer pipe are 1.8 m and 2.5 m, respectively. The spacing in longitudinal and circumferential directions is 0.5 m and 0.3 m, respectively. The outer layer grouting pipes were injected with cement grout, whereas the inner layer grouting pipes were injected with sand solidification agent. Grouting pressure was controlled strictly within a range from 0.8 Mpa to 1.0 MPa in case of ground uplifting.

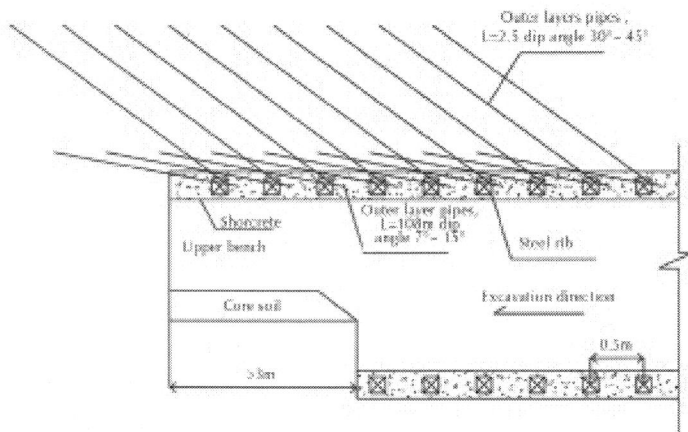

Figure 16: Double-layer grouting reinforcement.

The main purpose of this type of double-layer grouting activity was to improve, stabilize, and strengthen the strata prior to the excavation. The mechanical parameters of the region between the tunnel crown and ground surface were improved by injecting cement grouts, and an umbrella above the tunnel crown was produced. This was helpful for restricting the ground movement and minimizing the adverse effects on the surrounding environments.

Backfill Grouting behind Primary Lining

During construction, there are plenty voids appearing behind the primary lining, which were created inadvertently. These voids are located in the region between linings and surrounding ground. To prevent the ground movements induced by the voids, the $\varphi 42$ steel pipe, length of 0.6 m, was adopted to inject cement grout from inside of the pilot heading. Grouting pipes were arranged along the tunnel crown, and 2 m in longitudinal spacing.

ENVIRONMENTAL RESPONSE TO TUNNELING

Along the tunnel alignment, there are a number of surface buildings located around the Zhongjie subway station. During the construction stage, the safety of these buildings was one of the key problems encountered in this project. The monitoring scheme was designed to track building settlements and ground subsidence prior to the excavation (see Figures 9 and 17) to guarantee environmental safety.

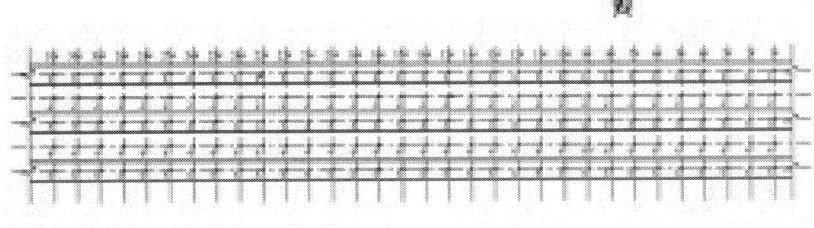

Figure 17: Plane view of the ground settlements monitoring points on a standard section.

The monitoring management standard of this project was classified into three grades: early warning value, alarm value, and limit value. Warning value is 70% that of the management critical value (shown in Table 2), whereas the alarm value and limit value are 85% and 100% that of the management critical value, respectively. The supporting patterns should be enhanced when the measured value reaches the early warning level. The excavation activities should be terminated immediately if the measured value reaches the limit level, and the evaluation of the safety situation of the surrounding buildings should be initiated.

Building Settlements

In this project, the settlement control indexes of the monitoring measurement value determined by the expert panel are shown in Table 3. The maximum settlement value for buildings was 40 mm. The settlement difference was controlled within 0.002 L (the length of existing building in the direction that is perpendicular to the tunnel axis). The control standard for the maximum ground surface settlement is 140 mm. Many shallow-buried projects in China have proven that the ground settlement allowable value should be controlled within 30 mm, which is unreasonable for a shallow-buried subway station.

Table 3: Management critical value for settlement of ground surface and building

Construction step	Surface settlement	Building settlements		Horizontal convergence	Crown settlement
		Absolute	Differential		
Pilot heading excavation	80	25	—	10	50
Supporting arch	35	10	—	—	20
Main structure	25	5	—	15	—
Summation (mm)	140	40	0.002	15	70

Settlement monitoring points were installed around each surface building. The settlements of these buildings were observed before, during, and after construction of the Zhongjie station. As shown in Figure 9, these buildings are located within the range of tunneling influence region. The settlements of the three buildings are shown in Figure 18. The maximum settlement of surface buildings induced by the excavation (including dewatering) was approximately 19.46 mm, which was observed with the Meigui Hotel. This is most likely because the Meigui Hotel is the highest building (20 stories, 70 m in height) among the affected surface buildings, and its foundation depth is relatively shallow (−7.5 m). Sirivachiraporn investigated the tunneling effects on buildings founded on different size pile lengths, and he stated that buildings on deep buried foundation displayed the least induced settlement [25]. In this project, the surface building that was least affected by the subway station excavation was the Shenyang commercial center, with a settlement of approximately 9.84 mm. This is mainly due to the structure having the deepest buried foundation depth (−11.6 m) among the affected buildings (see Table 2). The measured settlements of these buildings were controlled below 20 mm, and the deflection of these buildings was also smaller than 2 mm/m. This indicates that the risk mitigation measures had a significant effect on ground movement restriction. A further inspection of Figure 18 reveals that buildings can experience approximately 60% of their total settlement during the pilot heading excavation (from step 1 to step 3). This trend is similar to the numerical simulation presented in Section 3.2.

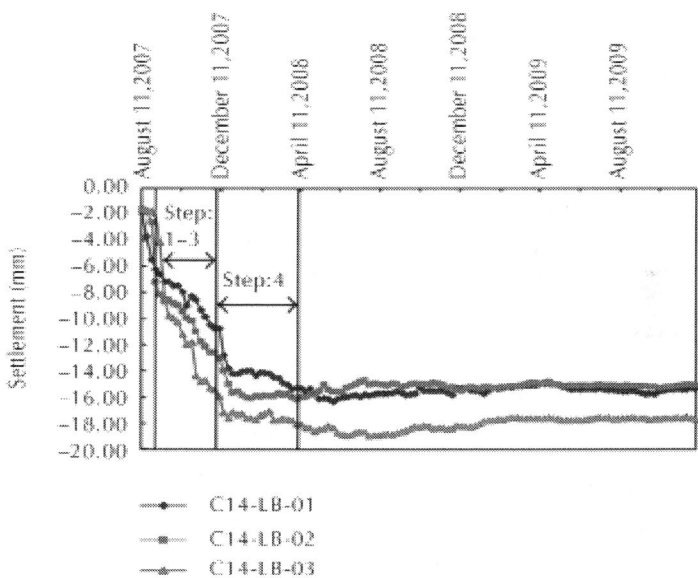

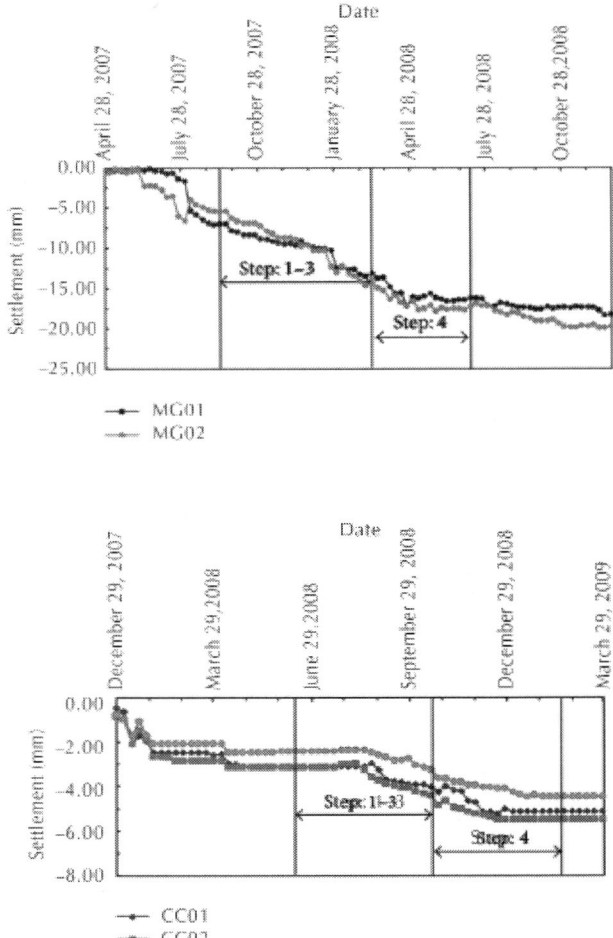

Figure 18: Long-term building settlement. (a) Laobian restaurant, (b) Meigui Hotel and (c) Shenyang commercial center.

Ground Surface Settlements

During construction, there are approximately 320 (some of them were destroyed during the construction) monitoring points installed on the ground surface along the tunnel longitudinal direction. The monitoring points of a standard section are shown in Figure 17.

The statistical data for ground surface subsidence caused by the subway station excavation in this project are shown in Figure 19. A further inspection of Figure 19 indicates that settlements generally exceeded 30 mm in 74.08% of the measurements. In addition, 6.64% of the measurements exceeded 80 mm. The maximum settlement was approximately 91.7 mm. The ground surface settlement for the measurements was successfully controlled below the early warning value, that is, 75% of that of the management critical value, which is equal to 105 mm.

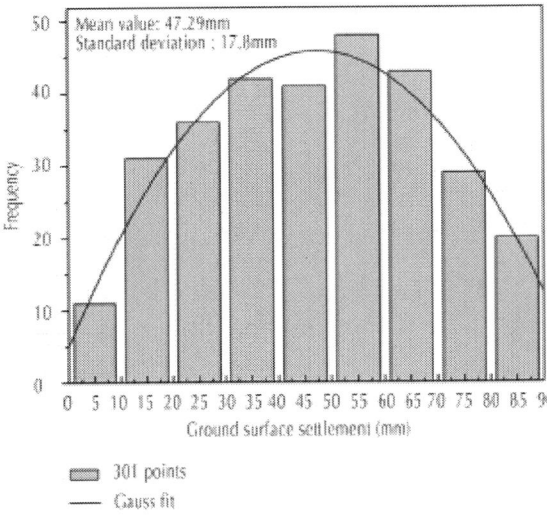

Figure 19: Statistical data of ground settlement induced by the subway station excavation.

Peck [26] reported that the transverse ground settlement profile can be expressed by a Gaussian distribution curve (see Figure 20). He derived the following empirical equation from their collected field measurement data.

$$S = S_{max} \exp\left(\frac{-y^3}{2i^2}\right),$$

(2)

Where s_{max} is the maximum ground surface settlement on the tunnel centerline, y is the horizontal distance from the tunnel centerline, and i is the horizontal distance from the tunnel centerline to the point of inflection on the settlement trough, which determines the shape and scope of the settlement trough.

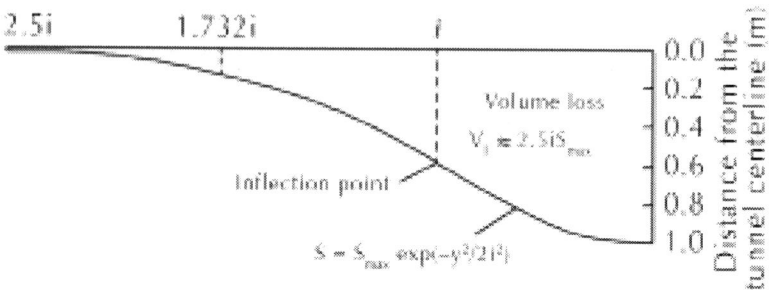

Figure 20: Transverse surface settlement trough curve.

The volume loss V_1 induced by tunneling can be obtained by integrating (2) along the distance y resulting in the following:

$$V_1 = \int_{-\infty}^{+\infty} Sdy = \sqrt{2\pi i S_{max}} \approx 2.5 i S_{max} . \tag{3}$$

O'Reilly and New [27] proposed a relationship between tunnel buried depth and , based on the monitoring data obtained from a UK tunneling project. The linear function can be expressed as follows:

$$i = Kz_0, \tag{4}$$

where z_0 is the tunnel buried depth and is the parameter of settlement trough, and its value is determined by stratigraphic condition.

The field observation data collected from 12 monitoring sections were fitted by a Gaussian distribution curve and are shown in Figure 21. It can be seen that the measured settlements are distributed in the area enclosed by the Gaussian fitting curves for values ranging from 9.65 m (upper bound) to 11.5 m (lower bound), corresponding to 0.66–0.82 of transverse settlement trough parameter . The tunnel volume losses derived from (3) are in the range from 0.36% to 0.85%. This value is relatively small compared to the findings of other research reported by Mair [28] for open face tunneling in stiff clays, which were usually between 1% and 2%. This difference is mainly due to the different ground conditions and tunnel excavation methods.

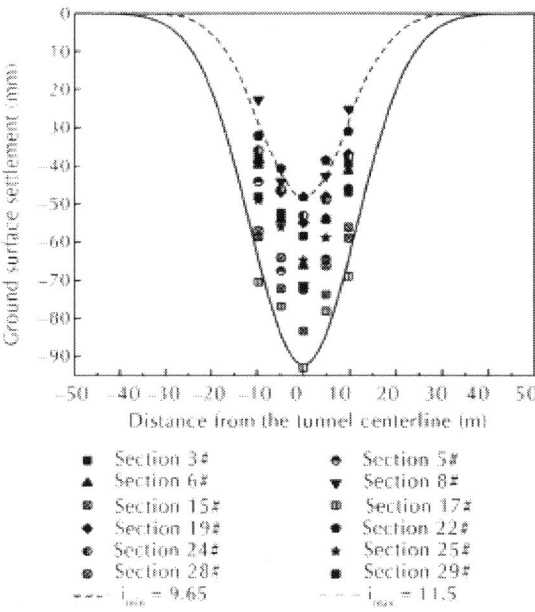

Figure 21: Fitting results of ground surface settlement trough.

ONCLUSIONS

The paper focused on the key techniques and risks management in a case study of the Zhongjie subway station, which was constructed in sandy soil using the PBA excavation method in a densely populated urban area. The main conclusions derived from the study are as follows. (1)The numerical simulation results indicate that. for a subway station constructed using the PBA excavation method, the excavation of pilot headings and the supporting arch constitute a high proportion of the ground surface settlement, 56.6% and 28.3% of the total settlement, respectively. Therefore, some risk mitigation measures should be undertaken during these excavation stages to restrict the ground movements and adverse impact on the surrounding environment. (2)Some risk mitigation measures were adopted in this project to ensure the environmental safety during the excavation, for example, appropriate groundwater dewatering scheme, feet-lock bolt, sleeve-pipe grouting technique, double-layer grouting pipe, and compensating grouting measures. These risk mitigation measures played a good role in ground movement control as no accidents occurred during or after the construction of the Zhongjie station.(3)After the construction was completed, the maximum deflection for the surface buildings was restricted below 2 mm/m, and the maximum surface building settlement was less than 20 mm. The surface building that was least affected by the subway station excavation was the Shenyang commercial center, which has a deep-buried foundation. The settlements of buildings on relatively shallow-buried foundation were larger than those of buildings on deep-buried foundation. The maximum ground settlement was 91.7 mm, which was also controlled below the early warning level, that is, equal to 105 mm.(4)Field observation data collected from 12 monitoring sections indicate that the transverse surface settlement measurements could subsequently be well fitted by a Gaussian distribution curve, and the settlement trough parameter was in the range of 0.66 to 0.82, which corresponds to 0.36%~0.85% of tunnel volume loss. The relatively low volume loss indicates the effectiveness of several types of grouting measures on ground movement restriction. (5)Monitoring during shallow-buried subway station construction plays a significant role in environmental risks management. It is helpful to implement construction risk assessments at each stage of construction and provide available information for the decisions that need to be taken or modified at each construction step.

CONFLICT OF INTERESTS

The authors declare that there is no conflict of interests regarding the publication of this paper.

ACKNOWLEDGMENTS

This paper is financially supported by the National Science and Technology Support Plan of 12th Five-Year Plan (2012BAK24B00, 2012BAK 24B0104) and the Fundamental Research Funds of Chinese Ministry of Education (N110301003).

REFERENCES

1. H. Chakeri, Y. Ozcelik, and B. Unver, "Effects of important factors on surface settlement prediction for metro tunnel excavated by EPB," Tunnelling and Underground Space Technology, vol. 36, pp. 14–23, 2013.

2. Y.-B. Luo and J.-X. Chen, "Mechanical characteristics and mechanical calculation model of tunnel feet-lock bolt in weak surrounding rock," Chinese Journal of Geotechnical Engineering, vol. 35, no. 8, pp. 1519–1525, 2013 (Chinese).

3. H. Huang, "State-of-the-art of the research on risk management in construction of tunnel and underground works," Chinese Journal of Underground Space and Engineering, vol. 2, no. 1, pp. 13–20, 2006 (Chinese).

4. A. Jurado, F. de Gaspari, V. Vilarrasa et al., "Probabilistic analysis of groundwater-related risks at subsurface excavation sites," Engineering Geology, vol. 125, pp. 35–44, 2012.

5. S. D. Eskesen, P. Tengborg, J. Kampmann, and T. H. Veicherts, "Guidelines for tunnelling risk management: International Tunnelling Association, Working Group no. 2," Tunnelling and Underground Space Technology, vol. 19, no. 3, pp. 217–237, 2004.

6. H. H. Einstein, "Risk and risk analysis in rock engineering," Tunnelling and Underground Space Technology, vol. 11, no. 2, pp. 141–155, 1996.

7. J. J. Reilly, "The management process for complex underground and tunneling projects," Tunnelling and Underground Space Technology, vol. 15, no. 1, pp. 31–44, 2000.

8.   K. You, Y. Park, and J. S. Lee, "Risk analysis for determination of a tunnel support pattern," Tunnelling and Underground Space Technology, vol. 20, no. 5, pp. 479–486, 2005.

9.   Q. Fang, D. Zhang, Y. Hou, B. Li, and F. Sun, "Safety risk control technology of urban subway with shallow tunnel construction method," Journal of Beijing Jiaotong University, vol. 34, no. 4, pp. 16–21, 2010.

10.   T. Wang, F.-R. Luo, W.-N. Liu, and X.-G. Li, "Influence of metro station construction by drift-pile-beam-arch method on soil and rigid-joint pipeline," Rock and Soil Mechanics, vol. 32, no. 8, pp. 2533–2538, 2011 (Chinese).

11.   T. Wang, W. Liu, C. Zhang, H. He, and X. Li, "Study on ground settlement induced by shallow metro station constructions," Chinese Journal of Rock Mechanics and Engineering, vol. 26, no. 9, pp. 1855–1861, 2007 (Chinese).

12.   T. Wang, F. Luo, W. Liu, and X. Li, "Study of surface settlement and flexible joint pipeline deformation induced by metro station construction with PBA method," China Civil Engineering Journal, vol. 45, no. 2, pp. 155–161, 2012 (Chinese).

13.   Y. Yang, L. Weining, D. Deyun, et al., "Analysis on heading excavation optimization in metro station constructed by drift-PBA method," Chinese Journal of Underground Space and Engineering, vol. 7, supplement 2, pp. 1692–1696, 2011 (Chinese).

14.   H.-J. He, Influence and control of metro tunnelling by drift-PBA method on adjacent bridge piles [Ph.D. thesis], Beijing Jiaotong University, 2007 (Chinese).

15.   Q. Fang, D. Zhang, and L. N. Y. Wong, "Shallow tunnelling method (STM) for subway station construction in soft ground," Tunnelling and Underground Space Technology, vol. 29, pp. 10–30, 2012.

16.   J.-F. Liu, T.-Y. Qi, and Z.-R. Wu, "Analysis of ground movement due to metro station driven with enlarging shield tunnels under building and its parameter sensitivity analysis," Chinese Journal of Underground Space and Engineering, vol. 28, no. 1, pp. 287–296, 2012 (Chinese).

17.   J. H. Shin, I. K. Lee, Y. H. Lee, and H. S. Shin, "Lessons from serial tunnel collapses during construction of the Seoul subway line 5,"

Tunnelling and Underground Space Technology, vol. 21, no. 3-4, pp. 296–297, 2006.

18. R. A. Forth, "Groundwater and geotechnical aspects of deep excavations in Hong Kong," Engineering Geology, vol. 72, no. 3-4, pp. 253–260, 2004.

19. M.-S. Wang, "Outline of tunnel construction by means of method of undercutting with shallow overburden," Tunnel Construction, vol. 26, no. 5, pp. 1–4, 2006 (Chinese).

20. M. Fraldi and F. Guarracino, "Limit analysis of collapse mechanisms in cavities and tunnels according to the Hoek-Brown failure criterion," International Journal of Rock Mechanics and Mining Sciences, vol. 46, no. 4, pp. 665–673, 2009.

21. M. Lei, L. Peng, and C. Shi, "Calculation of the surrounding rock pressure on a shallow buried tunnel using linear and nonlinear failure criteria," Automation in Construction, vol. 37, pp. 191–195, 2014.

22. M.-S. Wang, "The boring excavation method and construetion in Beijing metro," Chinese Journal of Rock Mechanics and Engineering, vol. 8, pp. 52–62, 1989 (Chinese).

23. Y. Xiang, S. He, M. Zhang, Z. Cui, and S. Ma, "Constraint effect of pilot-drift and separation-pile structure on ground movements induced by shallow tunneling," Chinese Journal of Rock Mechanics and Engineering, vol. 23, no. 19, pp. 3317–3323, 2004 (Chinese).

24. S.-C. Wu, A.-B. Jin, and Y.-T. Gao, "Studies of sleeve-valve-pipe grouting technique and its effect on soil reinforcement," Rock and Soil Mechanics, vol. 28, no. 7, pp. 1353–1358, 2007 (Chinese).

25. A. Sirivachiraporn and N. Phienwej, "Ground movements in EPB shield tunneling of Bangkok subway project and impacts on adjacent buildings," Tunnelling and Underground Space Technology, vol. 30, pp. 10–24, 2012.

26. R. B. Peck, "Deep excavation in soft ground," in Proceedings of the 7th International Conference on Soil Mechanics and Foundation Engineering, pp. 225–290, Mexico City, Mexico, 1969.

27. M. P. O'Reilly and B. M. New, "Settlements above tunnels in the United Kingdom—their magnitude and prediction," Tunnelling, pp. 173–181, 1982.

28. R. J. Mair, "Settlement effects of bored tunnels," in Proceedings of the International Symposium on Geotechnical Aspects of Underground Construction in Soft Ground, pp. 43–53, Balkema, London, UK, 1996.

Chapter 6

Effect of Sand Relative Density on Response of a Laterally Loaded Pile and Sand Deformation

Bingxiang Yuan,[1]Rui Chen,[1]Jun Teng,[1]Yixian Wang,[2]Wenwu Chen,[3]Tao Peng,[1]Zhongwen Feng,[1]Yang Yu,[4]andJianghui Dong[4]

[1]Shenzhen Graduate School, Harbin Institute of Technology, Shenzhen 518055, China

[2]School of Civil Engineering, Hefei University of Technology, Hefei 230009, China

[3]School of Civil Engineering and Mechanics, Lanzhou University, Lanzhou 730000, China

[4]School of Computer Science, Engineering and Mathematics, Flinders University,

ABSTRACT

Two scale-model tests were separately conducted in standard Toyoura sand with relative density of 50% and 80%. The effect of sand relative density on pile-soil interaction was investigated through the response of a laterally loaded pile and the sand movement around the pile. At a displacement of 3.6 mm of the loading point, the applied loads in loose and dense sand were 4.775 N and 21.025 N, respectively, and the maximum moment and soil resistance of the pile in dense sand were over 4 times those in loose sand. However, the deflection of the pile in dense sand was less than that in loose sand; additionally, the depth of zero deflection in dense sand was also less than that in loose sand. At the same time, the maximum displacements of loose sand in the vertical profile and ground surface were over 1.5 times those of dense sand. These characteristics occurred because the relative stiffness ratio of soil and pile increased as the relative density increased, which caused the behavior of the pile in dense sand to be elastic rather than rigid. In addition, the compacted sand particles did not move as easily as the loose sand particles.

INTRODUCTION

Many transmission towers, high-rise buildings, and bridges are supported by piles [1, 2]. These structures not only bear axial loads but also are subjected to considerable lateral loads such as violent winds and earthquakes [3]. Therefore, the lateral loading capacity of piles is an important design consideration for the construction of deep foundations [4]. Many approaches have been proposed to analyze the lateral loading capacity, such as the Broms method [5], the elastic method [6], the p-y curve approach [7, 8], and the strain wedge method [9, 10].

Most of research about the laterally loaded pile has been carried out by attaching strain gauges on piles to measure the lateral loading capacity, pile deflection, and soil resistance created by the pile [11, 12]. It is worth noting, however, that the behavior of the laterally loaded pile depends on soil reaction and vice versa. Only a few model tests had been used to study the movement of the soil around the pile. Of those, Otani et al. [13] used X-ray computed tomography (CT) to investigate three-dimensional deformation of the sand around a laterally loaded pile. Due to the high cost of the CT scanners, this technique has limited application in the geotechnical engineering. With the development of digital image processing, an economical, accurate, and full field image correlation technique, termed as particle image velocimetry (PIV), has been used in geotechnical engineering [14]. Liu et al. [15] and Yuan et al. [16, 17] used the PIV technique to correlate two consecutive images and measure the displacement fields of the ground surface around a pile under lateral loading.

This study presents the results from two scale-model tests done in dense and loose sand to determine the effects of relative density on the response of a laterally loaded pile and on the surrounding soil deformation. Combined with horizontal and vertical displacement fields measured using a PIV technique, the bending moment, lateral deflection, and soil resistance distribution along the model pile as derived from strain measurements were also analyzed. Our findings indicate that the effect of relative density on soil-pile interaction can be evaluated quantitatively.

EXPERIMENTAL SETUP AND TEST PROCEDURE

An Optical Experimental Setup

As shown in Figure 1, the optical experimental setup included a model box, a model pile, two cameras placed on the top and side of the model box, a stepping motor with a driver to apply lateral loading, and a data acquisition system to collect strain data.

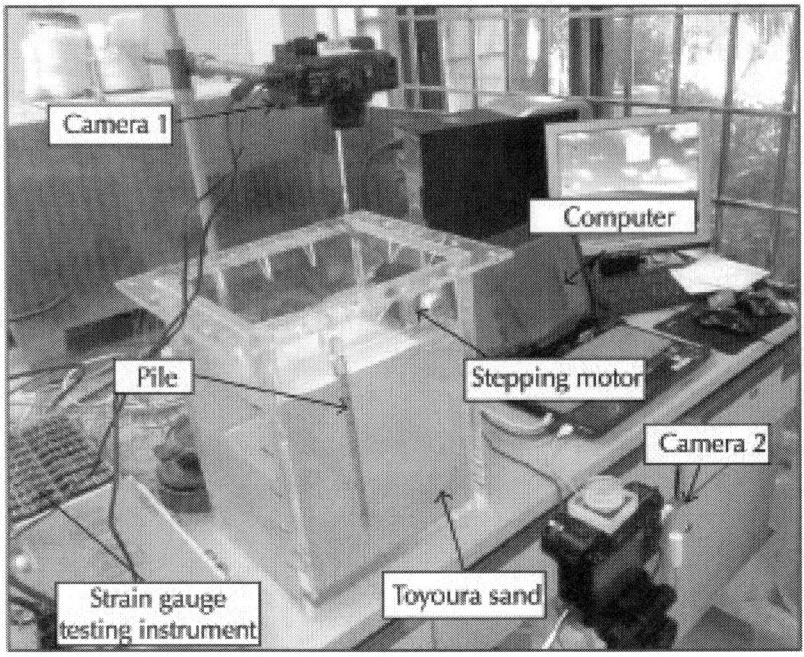

Figure 1: The optical experimental setup.

A Plexiglas box 300 mm in height, 200 mm in width, and 250 mm in length was used as the model box. The pile was a square section of Plexiglas with a width of 10 mm and a flexural rigidity of 50 N·m². The model pile was embedded 210 mm deep with a section extending approximately 40 mm above the surface. The lateral loads were applied using a stepping motor attached to the pile at a point located 20 mm below the pile head using wire. To investigate the response of the pile, seven pairs of strain gauges were attached along the pile. The readings from the stain gauges are obtained using a DH 3815 strain gauge testing instrument.

Two cameras (Canon PowerShotG10) with 4416 × 3312 pixels resolution were used to synchronously capture images by a developed software driver using MATLAB commands [18]. One camera was set in front of the model box with its optical axis perpendicular to the vertical profile, and another camera was set above the model box with its optical axis perpendicular to the ground surface, as shown in Figure 1. The built in camera zoom lenses were adjusted to select optimized regions of interest. While applying lateral loads, images of the vertical profile and ground surface were simultaneously captured by the two cameras. The images were employed to calculate the displacement fields of the vertical profile and ground surface using the PIV technique. More details about the PIV technique can be found in Liu and Iskander [19].

Soil Properties

The soil deposit was formed using uniform Toyoura sand. Over the past few decades, the properties and stress-strain behaviors of Toyoura sand have been extensively studied by means of diverse laboratory tests. The sand was uniformly fine sand consisting of subrounded to subangular particles. It had the minimum and maximum void ratios of 0.597 and 0.977, respectively, and a specific gravity of 2.65 [20]. The sand had uniformity coefficient of 1.7 and mean diameter of 0.17. The critical state friction angle, was 31° [21]. The loose and dense sand samples were prepared with relative density of 50% and 80%, respectively.

Test Results and Analysis

Two model tests of the laterally loaded pile were separately conducted in loose and dense sand. The effect of sand relative density on pile-soil interaction was investigated through the response of the laterally loaded pile and the sand movement around the pile.

Response of the Laterally Loaded Pile

The curves of applied load versus lateral displacement at the pile head in loose and dense sand are shown in Figure 2. Compared with the two curves, it was clearly found that the lateral bearing capacity in dense sand was larger than that in loose sand. When the lateral displacement was 3.6 mm (labeled in Figure 2), the applied loads in loose and dense sand were 4.775 N and 21.025 N, respectively. At this position the response of the laterally loaded pile and the sand displacements were simultaneously measured.

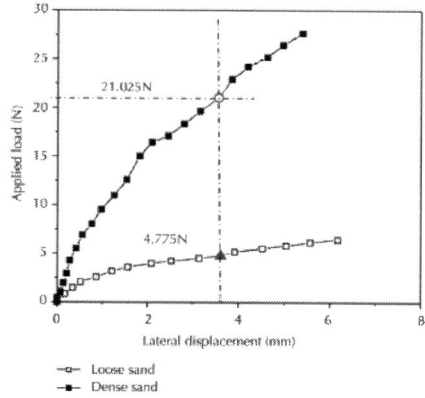

Figure 2: Curves of applied load versus lateral displacement in loose and dense sand.

The bending moments can be derived from the readings of the strain gauges attached along the model pile. Bending moment distributions along the pile were recorded for a lateral displacement of the pile head at 3.6 mm (labeled in Figure 2). As shown in Figure 3, the bending moments of the pile in dense sand were larger than those in loose sand. The maximum moment in dense sand was 4.2 times that in loose sand, and the maximum moments were both at the depth of 60 mm (i.e., 6 times pile width).

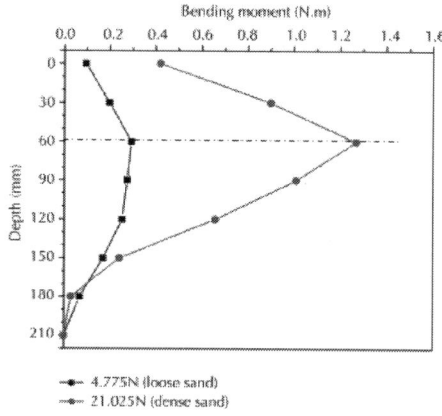

Figure 3: Comparison of the bending moment in loose and dense sand.

The lateral displacements of the model pile can be determined using a double integration method from the moment distribution function. As shown in Figure 4, the lateral displacements along the model pile in dense sand were less than those in loose sand. The depth of zero deflection in dense sand was also less than that in loose sand. The curve of the pile deflection in loose sand was a straight line; however, the pile in dense sand was flexural deformation, implying that the model pile in dense soil became an elastic pile rather than a rigid pile as occurred in loose sand.

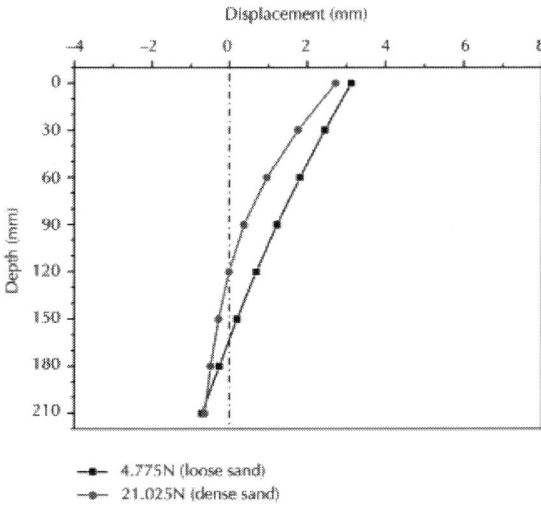

Figure 4: Comparison of the pile deflection in loose and dense sand.

The soil resistances in relation to depth can be interpreted with double differentiation of the bending moment distribution function. The trend of the soil resistances in dense sand was different from that in loose sand, as shown in Figure 5. The soil resistances along the pile in dense sand were larger than those in loose sand, and the maximum resistance in dense sand was 4.8 times that in loose sand, implying that the soil resistances increased with the increasing relative density. With the increasing relative density, the relative stiffness ratio of soil and pile increases, which causes the behavior of the pile to be as an elastic pile.

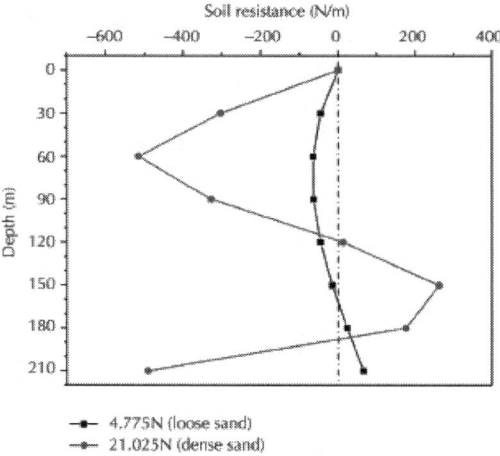

Figure 5: Comparison of the soil resistance in loose and dense sand.

The p-y curves of the pile in dense and loose sand can be obtained from the aforementioned calculation. As shown in Figure 6, the soil resistance increased as the lateral displacement increased. The increasing ratio of the soil resistance in dense sand was much larger than that in loose sand, implying that the soil resistance coefficient increased with the increasing relative density.

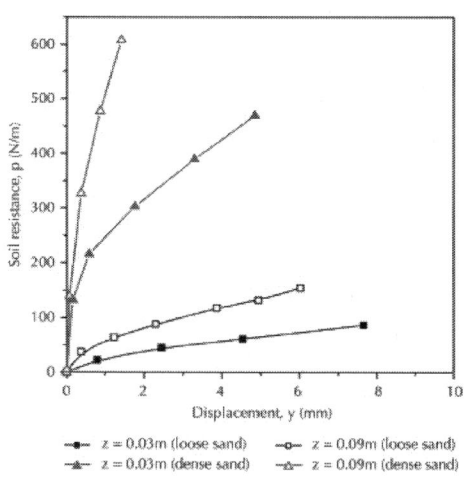

Figure 6: Comparison of p-y curves in loose and dense sand.

Soil Deformation around the Laterally Loaded Pile

The pile-soil interaction not only included the response of the laterally loaded pile, but also consisted of the deformation of the soil restraining the pile deflection. When the pile head was under the same lateral displacement of 3.6 mm in dense and loose sand, the displacement fields of the vertical profile and ground surface were calculated using the PIVview2C software. In the contours of the displacement fields at the vertical profile shown in Figure 7, the maximum displacement of loose sand was 1.5 times that of dense sand, and the influence zone along the depth in loose sand was obviously larger than that in dense sand. There were two main reasons: (1) the pile displacements in loose sand were larger than those in dense sand shown in Figure 4, and the pile in loose sand pushed the surrounding sand with a larger lateral displacement; (2) the dense sand was compacted; thus, the dense sand particles did not so easily move as the loose sand particles.

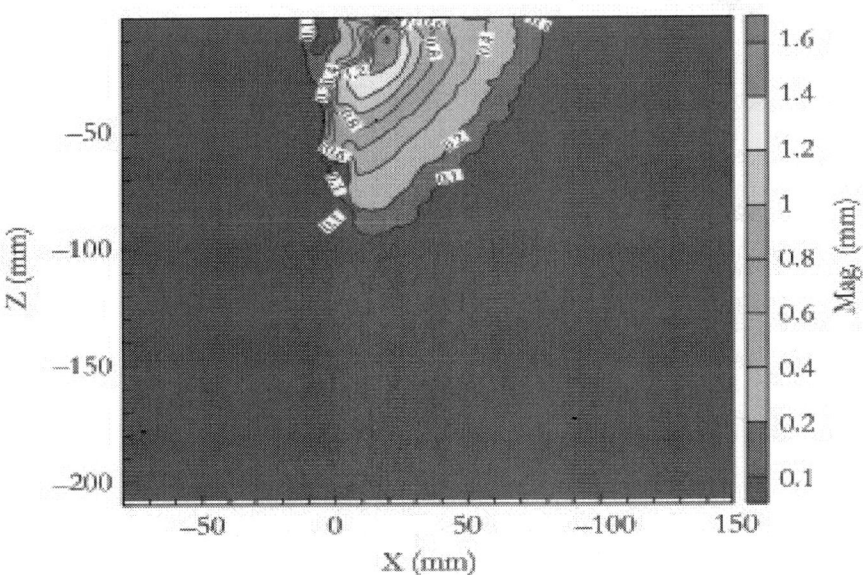

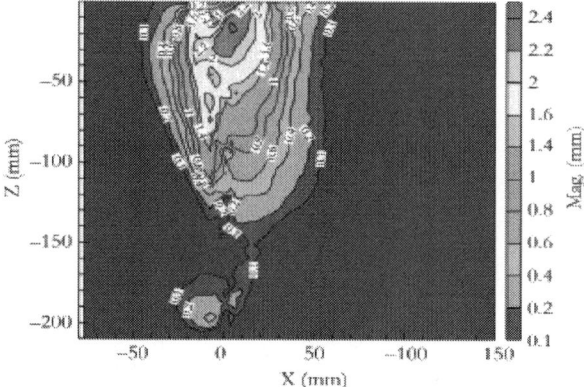

Figure 7: Contours of displacement fields at the vertical profile: (a) contours of displacement field in dense sand; (b) contours of displacement field in loose sand.

Comparing the contours of the displacement fields at the ground surface shown in Figure 8, the maximum displacements were 1.4 mm in dense sand and 2.4 mm in loose sand, which was consistent with the contours of the displacement fields at the vertical profile. It was also worth noting that the influence zone of the ground surface in loose sand was larger than that in dense sand, especially within the back of the pile. The loose sand particles easily moved, which caused the soil behind the pile to move downwards duo to diminished resistance of the pile. The compressed sand particles behind the pile were stable, even though losing the resistance of the pile, which was consistent with displacements at the vertical profile shown in Figure 7.

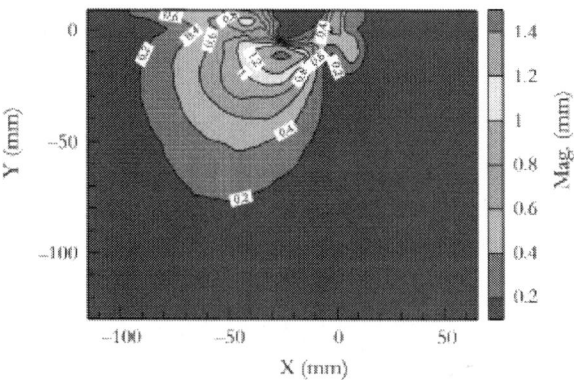

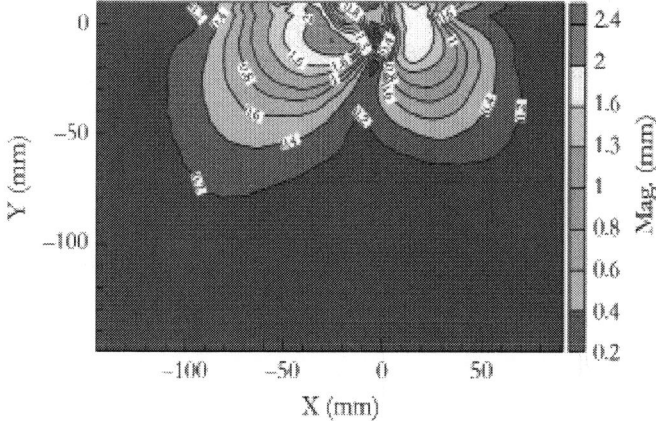

Figure 8: Contours of displacement fields at the ground surface: (a) contours of displacement field in dense sand; (b) contours of displacement field in loose sand.

CONCLUSIONS

In this study, two scale-model tests in dense and loose sand were conducted to investigate the effect of relative density on the response of a laterally loaded pile and surrounding soil deformation. The bending moment, lateral deflection, and soil resistance distribution along the model pile were derived from the strain measurement. When the same displacement of 3.6 mm at the loading point of the pile head was reached, the lateral bearing capacity, bending moment, and soil resistance in dense sand were larger than those in loose sand; however, the deflection of the pile in dense sand was less than that in loose sand. The main reason was that the relative stiffness ratio of soil and pile increased with increasing relative density, which caused the behavior of the pile in dense sand to be elastic rather than rigid.

The displacement fields around the laterally loaded pile in the vertical profile and ground surface were simultaneously measured using the PIV technique. At the same displacement of the pile head, the maximum displacements of loose sand in the vertical profile and ground surface were over 1.5 times those of dense sand. In addition, the influence zones along the depth of the pile and the back of the pile in loose sand were obviously larger than those in dense sand. The main reasons were that the pile displacements in loose sand were larger than those in dense sand, the pile in loose sand pushed the surrounding sand with a larger displacement, and the dense sand was compacted; thus, the dense sand particles did not so easily move as the loose sand particles, and the sand particles behind the pile were stable, even though diminishing resistance of the pile.

The results demonstrated that the combined analysis of the response of the pile and the sand displacement fields is appropriate to investigate the mechanism of the effect of relative density on pile-soil interaction. In the future research, the internal sand movement will be observed using transparent soil to investigate the effect of the sand density on the internal pile-soil interaction.

CONFLICT OF INTERESTS

The authors declared that they have no conflict of interests to this work.

ACKNOWLEDGMENTS

The authors would gratefully like to acknowledge the support provided by the National Natural Science Foundation of China (no. 51308164, no. 51279049, and no. 51304057) and the China Postdoctoral Science Foundation (no. 2013M530157 and no. 2014T70349). The editorial help from Professor Galen Leonhardy of Black Hawk College is also greatly appreciated.

REFERENCES

1.   D. Leshchinsky, F. Vahedifard, and C. L. Meehan, "Application of a hydraulic gradient technique for modeling the uplift behavior of piles in sand," Geotechnical Testing Journal, vol. 35, no. 3, pp. 400–408, 2012.

2.   S. K. Suryasentana and B. M. Lehane, "Numerical derivation of CPT-based p-y curves for piles in sand,"Geotechnique, vol. 64, no. 3, pp. 186–194, 2014.

3.   A. M. A. Nasr, "Experimental and theoretical studies of laterally loaded finned piles in sand," Canadian Geotechnical Journal, vol. 51, no. 4, pp. 381–393, 2014.

4.   L. Ping, Z. Dong, W. Luo, and G. Qian, "The pressure-field characteristics around porous wind fences: results of a wind tunnel study," Environmental Earth Sciences, vol. 68, no. 4, pp. 947–953, 2013.

5.   B. B. Broms, "Lateral resistance of piles in cohesiveless soils," Journal of the Soil Mechanics and Foundations Division, vol. 90, no. 3, pp. 123–156, 1964.

6.   H. G. Poulos, "Behavior of laterally loaded piles: I-single piles," Journal of the Soil Mechanics and Foundations Division, vol. 97, no. 5, pp. 711–731, 1971.

7.   M. Ashour and H. Ardalan, "p-y curve and lateral response of piles in fully liquefied sands," Canadian Geotechnical Journal, vol. 49, no. 6, pp. 633–650, 2012.

8.   M. Khari, K. A. Kassim, and A. Adnan, "Development of p-y curves of laterally loaded piles in cohesionless soil," The Scientific World Journal, vol. 2014, Article ID 917174, 8 pages, 2014.

9.   M. Ashour, G. Norris, and P. Pilling, "Lateral loading of a pile in layered soil using the strain wedge model," Journal of Geotechnical and Geoenvironmental Engineering, vol. 124, no. 4, pp. 303–314, 1998.

10.   L.-Y. Xu, F. Cai, G.-X. Wang, and K. Ugai, "Nonlinear analysis of laterally loaded single piles in sand using modified strain wedge model," Computers and Geotechnics, vol. 51, pp. 60–71, 2013.

11.   D. Su, "Resistance of short, stiff piles to multidirectional lateral loadings," Geotechnical Testing Journal, vol. 35, no. 2, pp. 313–329, 2012.

12.   J.-S. Chiou, W.-L. Tai, and C.-H. Chen, "Lateral hysteretic behavior of an aluminum model pile in saturated loose sand," Journal of the Chinese Institute of Engineers, Transactions of the Chinese Institute of Engineers, Series A, vol. 37, no. 3, pp. 313–324, 2014.

13.   J. Otani, K. D. Pham, and J. Sano, "Investigation of failure patterns in sand due to laterally loaded pile using X-ray CT," Soils and Foundations, vol. 46, no. 4, pp. 529–535, 2006.

14. M. Ahmed and M. Iskander, "Analysis of tunneling-induced ground movements using transparent soil models," Journal of Geotechnical and Geoenvironmental Engineering, vol. 137, no. 5, pp. 525–535, 2011.

15. J. Liu, B. Yuan, V. T. Mai, and R. Dimaano, "Optical measurement of sand deformation around a laterally loaded pile," Journal of Testing and Evaluation, vol. 39, no. 5, pp. 754–759, 2011.

16. B.-X. Yuan, W.-W. Chen, T. Jiang, Y.-X. Wang, and K.-P. Chen, "Stereo particle image velocimetry measurement of 3D soil deformation around laterally loaded pile in sand," Journal of Central South University, vol. 20, no. 3, pp. 791–798, 2013.

17. B. Yuan, R. Chen, J. Teng, T. Peng, and Z. Feng, "Effect of passive pile on 3D ground deformation and on active pile response," The Scientific World Journal, vol. 2014, Article ID 904186, 6 pages, 2014.

18. B. Yuan, J. Liu, W. Chen, and K. Xia, "Development of a robust Stereo-PIV system for 3-D soil deformation measurement," Journal of Testing and Evaluation, vol. 40, no. 2, pp. 256–264, 2012.

19. J. Liu and M. Iskander, "Adaptive cross correlation for imaging displacements in soils," Journal of Computing in Civil Engineering, vol. 18, no. 1, pp. 46–57, 2004.

20. M. Yoshimine, K. Ishihara, and W. Vargas, "Effects of principal stress direction and intermediate principal stress on undrained shear behavior of sand," Soils and Foundations, vol. 38, no. 3, pp. 179–188, 1998.

21. R. Verdugo and K. Ishihara, "The steady state of sandy soils," Soils and Foundations, vol. 36, no. 2, pp. 81–91, 1996.

Simplified Analysis of Laterally Loaded Pile Groups

F.M. Abdrabbo and K.E. Gaaver

Structural Engineering Department, Faculty of Engineering, Alexandria University, Egypt

ABSTRACT

The response of laterally loaded pile groups is a complicated soil–structure interaction problem. Although fairly reliable methods are developed to predicate the lateral behavior of single piles, the lateral response of pile groups has attracted less attention due to the required high cost and complication implication. This study presents a simplified method to analyze laterally loaded pile groups. The proposed method implements *p*-multiplier factors in combination with the horizontal modulus of subgrade reaction. Shadowing effects in closely spaced piles in a group were taken into consideration. It is proven that laterally loaded piles embedded in sand can be analyzed within the working load range assuming a linear relationship between lateral load and lateral displacement. The proposed method estimates the distribution of lateral loads among piles in a pile group and predicts the safe design lateral load of a pile group. The benefit of the proposed method is in its simplicity for the preliminary design stage with a little computational effort.

INTRODUCTION

Lateral response of piled foundations is important in design of structures that may be subjected to lateral loads. The lateral loads acting on piled foundations may be sustained, as earth pressure on a retaining wall, or alternated, as from reciprocating machinery, or pulsated, as from the traffic loading on a bridge pier. Lateral loads are in the order of 10–15% of the vertical loads in case of onshore structures, while this value may exceed 30% in case of offshore structures [22]. The response of a laterally loaded pile is a complicated soil–structure interaction problem; because pile deflection depends on soil reaction and in turn soil reaction influences by pile deflection. Fairly reliable methods have been developed for predicting the lateral response of single piles, since the pioneer works of Matlock and Reese [15], and Broms [6]. Frechette et al. [10] reviewed the design methods for laterally loaded groups of drilled shafts and compared between methods employing a group reduction factor and a *p*-multiplier. Kumar and Lalvani [13] analyzed the nonlinear load–deflection behavior of laterally loaded piles using *p–y* relationships. Full scale and centrifuge model tests on pile groups have been conducted by Brown et al. [5], McVay et al. [17], and Rollins et al. [28].

Laterally loaded pile groups may be analyzed using elastic continuum approach [21], and group equivalent pile procedure [19]. The *p–y* relationships, initially developed by Matlock [16], have been used to model pile–soil interaction [24]. As a result of the interaction between piles in a group, *p–y* relationship of single pile was modified to be implemented in pile group analysis. The modifications can be carried out by introducing p-multiplier [19] and [28]. The p-multiplier concept is an effective procedure for implementing in pile group analysis; however unique values of p-multiplier for a pile group are not standardized. It is important to note that p–y relationship is not a soil property, but rather pile–soil property [1].

In order to provide simple solutions to complex problems such as lateral response of pile groups, various simplifications are reported [31]. Some practical methods are based on trial and adjustment processes, starting with a very simple approach to obtain an approximate response. The process can then be elaborated to some degree to narrow the margin of error. Very elaborate calculation processes are not justified, because of the non-homogeneity of most natural soil deposits and the soil disturbance caused by installing process of piles. In recent years, several simplified approaches for analysis of laterally loaded single piles or pile groups have been developed that can be used with little computational effort [14] and [7]. This paper presents a proposed simplified method, for analyzing laterally-loaded pile groups, using p-multipliers in combination with Winkler's model.

GEOTECHNICAL DATA OF THE SITE

In a construction site located at the Northeast of Nile River Delta, Damietta free trade district, Egypt, lateral loading tests on vertical single piles were conducted. These tests were done to confirm the design lateral working load of the constructed piles. But the outputs of these tests cannot be used directly to verify the design lateral load of the pile groups. The proposed method was implemented to analyze the lateral behavior of pile groups based on the results of lateral loading tests on single piles. Before discussion of the proposed method, the obtained soil stratifications and soil properties from geotechnical investigation are presented.

Forty-eight mechanical boreholes were conducted at the construction site up to 60 m depth. The retrieved soil samples from the boreholes were classified in accordance with ASTM D 2487. Fig. 1 illustrates the soil profile through a typical borehole at the construction site. The soil profile consists of a top layer of medium dense sand up to a depth of 15 m. This layer is underlain by a thick layer of soft to medium normally consolidated clay, which is extending up to a depth of 36 m. At this depth, a very dense sand bed was encountered and explored to a depth of 60 m. The ground water table is at 1.0 m below the ground surface. The shear strength parameters of the top sand layer were interpreted from the average value of SPT. So, the top sand has an average natural unit weight of 18 kN/m³, angle of internal friction 37°, and relative density of 70%. Moreover, the parameters of shear strength for clay and bottom sand layers were measured using triaxial test apparatus. The obtained soil parameters are shown in Fig. 1.

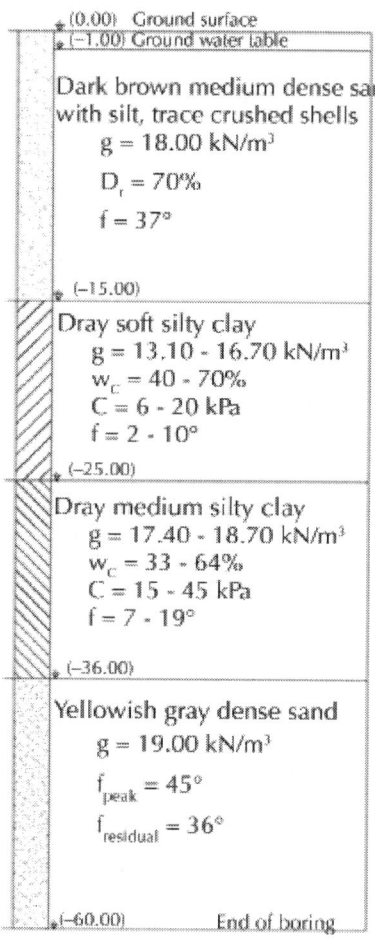

Figure 1: Typical borehole in Damietta site.

Bored piles, of 600 mm diameter, were constructed to be seated at a depth of 40 m below the existing ground surface. Each pile is reinforced by nine bars of 18 mm diameter of high tensile steel, grade 36/52. The pile reinforcement is extending to 16 m below the ground surface. Pile groups are attached to pile caps which are resting on the ground surface. As the pile is considered a flexible pile, the safe design lateral load of the pile depends on structural capacity of the pile cross section and the allowable lateral deflection at the pile head. Based on these design criteria, the safe design lateral load of single pile is 80 kN, dominated by structural capacity of the pile cross section.

The Proposed Method

The aim of the proposed method is to estimate the distributions of lateral load among piles in a group. The pile behavior under lateral loads depends upon the lateral stiffness of soil and the pile stiffness. Poulos and Davis [21] showed that the lateral stiffness of cohesionless soils increases linearly with depth. In the analysis, the horizontal subgrade reaction $(K_{x(s)})$ at a depth (Z) below the ground surface within the top sand layer is expressed as; $K_{x(s)} = _h \cdot Z$, where $_h$ is the modulus of horizontal subgrade reaction of top sand layer. The value of $_h$ was considered equal to 16.346 MN/m^3 for a single pile in the top sand layer. Reese and Matlock [25] suggested that the horizontal subgrade reaction through cohesive soils should be constant with depth. Also, Poulos and Davis [21] showed that the horizontal subgrade reaction of clays does not change with depth. Thus, a constant value of horizontal subgrade reaction through the soft silty clay layer $(K_{x(c)})$ is considered. It is significant to mention that modulus of horizontal subgrade reaction is not a unique soil property, but depends on pile characteristics and the lateral displacement of the pile.

The relative stiffness factor (T) and the maximum value of depth coefficient (Z_{max}) were calculated as; $T = (E \cdot I / _h)^{0.2}$, and $Z_{max} = (L_{ef}/T)$, where $E \cdot I$ is the flexural rigidity of the pile as un-cracked section. The maximum value of depth coefficient (Z_{max}) was found to be about 6.36, which indicates that the pile behavior is as a flexible pile. So, the piles in a group are considered flexible piles embedded in a stratified soil. Hence the lateral load may be resisted by soil lateral stresses developed along the top portion of the pile, which is called the effective pile length (L_{ef}). One method of assessing the value of (L_{ef}) is by modeling a single pile as a beam in a soil. The soil is represented by an elastic uncoupled spring modulus. L_{ef} is assessed as the depth where the lateral deflection of the pile is effectively zero. It is important to note that, the effective depth of a single pile differs from the effective depth of a pile in a group, due to interaction between piles in a group. The effective length (L_{ef}) of a pile in a group was calculated by re-analyzing single pile but with softer springs. These spring moduli were obtained by multiplying the spring modulus of the single pile by p-multiplier values. This means that piles in a group have different effective lengths. It is expected that as p-multiplier of the leading pile is bigger than p-multiplier of the trailing pile, the effective length of the leading pile is shorter than that of the trailing pile. It was established that halving the ($K_{x(s)}$) value of the top sand layer produces insignificant change in the response of laterally loaded single piles. Therefore there is no appreciable difference of the effective length of the piles within a group. Reese and Matlock [25] pointed out that the accuracy of horizontal subgrade reaction (k) is not critical. A 32 to 1 variation in (k) is required to produce 2 to 1 variation in the resulted bending moment. However, it should be realized that the value of (k) is essential empirical. But it may be vary with pile type, pile diameter, pile deflection, type of loading, and rate of loading.

Using the aforementioned soil profile and soil properties at Demiatta free trade district, the effective depth of single pile and for a pile within a group was founded to be about 16 times the pile diameter. Clearly, the effective depth is less than the depth of the top sand layer. So the lateral response of the piles is governed by the properties of top sand layer.

Dimensionless relationships, developed by Reese and Matlock [25], were used to determine the distribution of pile displacements, bending moments, shearing forces, soil resistances, and slope deflections along the effective length of a single pile. The safe design lateral load of 80 kN is considered at the pile head assuming fixed head piles, without any free length above the ground surface. For a single pile, $_h$ was implemented directly, while for a pile within a group $_h$ was reduced due to shadowing effects. Consider a pile group configuration containing n-rows and m-columns of piles, the lateral load is applied in a direction parallel to the row direction. The shadowing effect depends upon the location of pile column within the group and the location of the pile within the column. McVay et al. [17] concluded that in the same pile column, the middle pile develops slightly less lateral resistance than the side piles because it is subjected to more substantial shadowing effects. However, the authors showed that the difference is not significant and no significant error is developed by assuming that all piles in the same column carry the same lateral load. Consequently, the multiplier factor (p) for all piles within a column was assumed to be the same value.

To consider the effects of pile–soil–pile interaction in a group, one identical row of piles is considered. The lateral load is applied in a direction parallel to the row direction. A pile within a group was analyzed as a single pile and the straining actions along the pile were assessed for different values of $_{hp}$. The obtained straining actions include the distribution of pile deflections, bending moments, shearing forces, lateral soil resistances, and slope deflections. The reduced value of horizontal subgrade reaction ($_{hp}$) was obtained as; $_{hp} = p \cdot _{h}$. The values of p-multiplier factor were obtained from McVay et al. [17], and Ooi et al. [19]. As a result, a database containing pile deflections, bending moments, shearing forces, lateral soil resistances, and slope deflections were formed for a single pile embedded in fictitious sand of different values of ($_{hp}$) and subjected to different lateral loads. The database was formed with the help of computer spreadsheets.

Analysis Methodology

Once the database was compiled, the analysis of a laterally loaded pile group can be carried out. Two case studies are considered in the analysis. In the first case study, a pile group configuration containing $n \times m$ piles and subjected to certain lateral load (P_H) at the ground surface is considered. The lateral load is applied in a direction parallel to the row direction. The piles in the group are attached to a rigid pile cap, that is to say the lateral displacements of all piles in the group at their heads are equal. The properties and reinforcement of the piles are mentioned in Section 2. The lateral load contributed by friction between pile cap and soil is disregarded. So, the acting lateral load is applied on piles in the group. The unknowns in this case study are the load distribution among piles in the group and the lateral displacement of the group at ground surface. The piles are considered long flexible piles. The relationship between lateral applied load (P_H) and lateral displacement at the pile head (y_G) is assumed to be linear. The elasticity approach for analysis of laterally loaded piles was considered by Poulos

[20], and Sogge [30]. Furthermore, this assumption will be verified in the forthcoming sections.

Consider one identical row of piles, the load distribution among piles in a group and the lateral displacement of the group can be calculated as shown in the following steps:

1. The p-multiplier for each pile was assessed from documented literature. The leading pile has a bigger p-multiplier while the trailing pile has a smaller p-multiplier.

2. Entering an assumed value of lateral displacement (y_G) and the p-multiplier into the database, the lateral load acting on a pile (P_i) in each column corresponding to the assumed lateral displacement of the group (y_G) and the given specified p-multiplier was obtained.

3. If the sum of pile lateral loads (ΣP_i) is equal to the applied lateral load (P_H) acting on the pile group, the solution is obtained and the process is terminated.

4. If the sum of pile lateral loads (ΣP_i) differs from the applied lateral load (P_H), the assumed pile group displacement needs to be altered. This can be achieved by applying a correction factor to both assumed pile group displacement and loads on piles within the group. This correction factor is equal the ratio between the resulted (ΣP_i) and the applied lateral load (P_H).

5. The resulted lateral displacement after correction is the lateral displacement of the pile group.

Once the correction is achieved, the distribution of bending moment, shear force, and soil pressure in each pile in the group can be obtained from the collected database.

In the second case study, the number of piles in a group is known and the safe design lateral load of the single pile is also known. It is required to determine the safe design lateral load of the pile group. This case study represents a practical case in which the safe design lateral load of single pile is evaluated and verified by field loading tests. Usually the pile group configurations are assessed by knowing the vertical applied loads, vertical working load of single pile, and group efficiency. Once the pile groups are arranged, the capability of pile groups to sustain lateral loads safely becomes essential. To tackle this problem, the value of p-multiplier for each pile in the group was assessed. The properties and reinforcement of the piles are mentioned in Section 2. Consider one identical row of piles, the safe design lateral load of a pile group can be calculated as shown in the following four steps:

1. Using the compiled database, the head displacement of the leading pile (y_1) was obtained corresponding to horizontal subgrade reaction of $(p_1 \cdot \eta_{hp})$ and lateral applied load equal to design lateral load of the single pile.

2. At the same displacement (y_1) and horizontal subgrade reaction of $(p_2 \cdot \eta_{hp})$, the pile load for the second pile is obtained.

3. The procedure is repeated for all piles in a row of the group.

4. The design lateral load of a pile group is equal to the sum of individual pile loads within the pile group.

The only drawback of the proposed method is that the load of the leading piles is assumed equal to the design lateral load of single pile but with a corresponding bigger displacement compared to the displacement of individual pile. The p-multiplier of leading pile varies from 0.75 to 1.00 [28], and from 0.65 to 1.00 [19]. Truly at the same lateral displacement of a pile group and single individual pile, the lateral load applied on the leading pile in the group is smaller than single individual pile.

NUMERICAL EXAMPLE

Consider a group of three piles, each of 600 mm diameter installed in one row. This arrangement represents $n \times m$ pile groups, where $n = 1$, 2, 3, etc. and $m = 3$, Fig. 2. Single vertical pile was analyzed under a lateral load of 80 kN using relationships developed by Reese and Matlock [25]. Fixed head pile was assumed, the resulted pile head displacement is 1.95 mm. It is required to determine the lateral load acting on each pile in the group under an acting lateral load of 240 kN. The analysis was started by considering the p-multipliers as 0.8, 0.4 and 0.3 for leading pile, middle pile, and trailing pile respectively. These values were obtained by judgment from values recommended by McVay et al. [17], Dodagoudar et al. [8], Rollins et al. [29], Brown et al. [4], Ilyas et al. [12], Reese et al. [27], Mokwa and Duncan [18], and FHWA [9]. The reported data are for piles in all types of soil. The above cited literature indicated that, the p-multiplier for leading-column piles varies from 0.60 to 0.93 with an average value of 0.79. The *p*-multiplier for the second-column piles varies from 0.40 to 0.78 with an average value of 0.58. For the third-column piles, the *p*-multiplier varies from 0.40 to 0.63 with an average value of 0.46. For the fourth-column piles, the *p*-multiplier varies from 0.40 to 0.68 with an average value of 0.52. The reported values by Dodagoudar et al. [8] excluded values published by McVay et al. [17], which the present analysis was considered. Also the average value of *p*-multiplier reported by Dodagoudar et al. [8] for the fourth-column piles is bigger than the third-column piles. It is really difficult to consider the average values of p- multiplier due to different conditions of interpreted values. Also, one should realize that *p*-multiplier depends upon the value of (L/T).

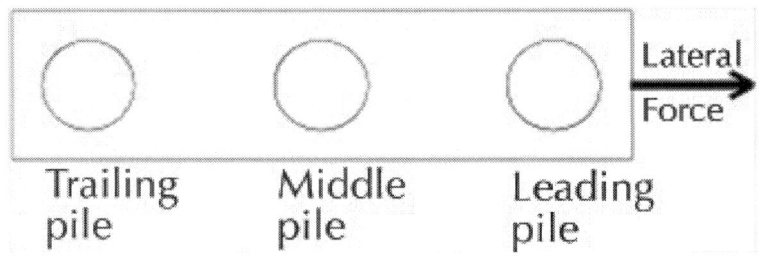

Figure 2: Pile arrangement, pile diameter = 0.60 m and length = 40 m.

The pile head lateral displacements of the piles under an applied lateral load of 80 kN were assessed from the compiled database. The values of $_{hp}$ equal to 0.8 $_{h'}$ 0.4 $_h$ and 0.3 $_h$ were considered for leading, middle, and trailing piles respectively. Accordingly, the head lateral displacements of leading, middle, and trailing piles are 2.23, 3.40, and 3.98 mm respectively. These lateral displacements violate the boundary conditions at the pile heads. Therefore by assuming that lateral deflection of the pile group at ground surface is 3.00 mm and each pile in the group exhibits this deflection, the lateral loads of leading pile, middle pile, and trailing pile are 107.62 kN, 70.58 kN, and 60.3 kN respectively. At this stage, the sum of the lateral loads of the piles in the group is 238.5 kN which is less than the applied lateral load of 240 kN. So the pile group displacement should be corrected to be 3.019 mm to match with the applied load of 240 kN. The corresponding acting loads on leading, middle, and trailing piles become 108.28 kN, 71.04 kN and 60.68 kN respectively. The sum of the pile loads in the group becomes equal to the applied lateral load. The corresponding lateral displacement of the group is 1.548 times the lateral displacement of the single pile.

It is worth noting that 9-pile group arranged in a square pattern exhibits the same lateral displacement under an acting lateral load of three times of 240 kN. The shadowing approach on which the piles in a row have no interaction effects on piles outside this row is contradicted with the elastic approach. In the elastic approach, the interaction factor between two piles depends upon the angle in plan between the centers of these two piles among other many factors [21]. Also, Randolph [23] suggested an expression to estimate the interaction factors between fixed-head piles in a group. On the other hand, limitations of the elastic approach to pile interaction under lateral loading had been revealed from model tests on a centrifuge carried by Barton [3].

To estimate the safe design lateral load of (1 × 3) pile group, the lateral load of the leading pile was assumed equal to the safe design lateral load of the single pile, which is 80 kN. From the compiled database and introducing p-multiplier of 0.80, the corresponding lateral displacement at the pile head of the leading pile is 2.23 mm. By enforcing the piles in the group to exhibit the same deflection, thus the lateral loads of the middle and the trailing piles become equal to 52.49 kN and 44.83 kN respectively. These values were obtained from the database using p-multipliers for middle and trailing piles. The maximum bending moments induced in leading, middle, and trailing piles under lateral loads of 80, 52.49, and 44.83 kN respectively and corresponding to p-multipliers of 0.80, 0.40, and 0.30 were obtained from the compiled database. The structural capacity, expressed as the bending capacity of the pile cross section, was calculated and compared with induced values. It was found that the pile cross section is capable to resist the induced moments safely. If the pile cross section is incapable to resist the induced bending moment, the process is repeated but with a small value of lateral load on the leading pile. In flexible piles, the dominant factor in assessing the safe lateral load of single pile and pile group is the structural capacity of piles. Consequently the safe design lateral load of the pile group is 177.32 kN.

Nine-pile group arranged in a square pattern carries three times the achieved value of lateral load, at the same value of lateral displacement.

The corresponding group reduction factor is 0.739, compared by 0.67 that was reported by Frechette et al. [10]. This analysis indicated that the leading pile carries 45.1% of the applied lateral load acting on the pile group, while the middle and trailing piles carry 29.6% and 25.3% respectively. The corresponding lateral displacement of the group is 1.143 times the lateral displacement of a single pile. McVay et al. [17] conducted lateral tests on pile groups founded in sand in a centrifuge machine. Their results indicated that, for 3 × 3 pile group the percentage of lateral load carried by leading, middle, and trailing piles were 43.3%, 31.5%, and 25.2% respectively in case of dense sand. For loose sand, the corresponding values were 46.6%, 29.3%, and 24.1% respectively. A comparison between the results of pile load distribution obtained by the proposed simplified method and the measured values revealed that results of the proposed method are in good agreement with the experimental results.

The only shortcoming of the proposed method is that the effect of spacing between piles in the group is not considered. However, from a practical point of view, most of designers prefer to arrange the piles at minimum spacing in a group in order to minimize the size of the pile cap. Thus the proposed method is suitable to be implemented for pile groups having practical spacing of 2.5–3 times the pile diameter. In this situation it is important to note that the effect of spacing between piles on the lateral response of a pile group can be considered, if p-multiplier database is developed for pile groups of different spacing.

Justification of the Assumptions

The proposed method is based on linear relationship between lateral load and lateral displacement at pile head, which was confirmed by McVay et al. [17], Yang and Liang [32], Gaaver [11], and field test results presented in Figure 3 and Figure 4. Fig. 3 presents results of a test pile of 600 mm diameter and 40 m length installed at the site located at Northeast side of Nile River delta, having the same succession of soil strata as given before in Section 2. Fig. 3 illustrates a good agreement between theoretical p–y relationship and the experimental values up to the design lateral load of 80 kN.

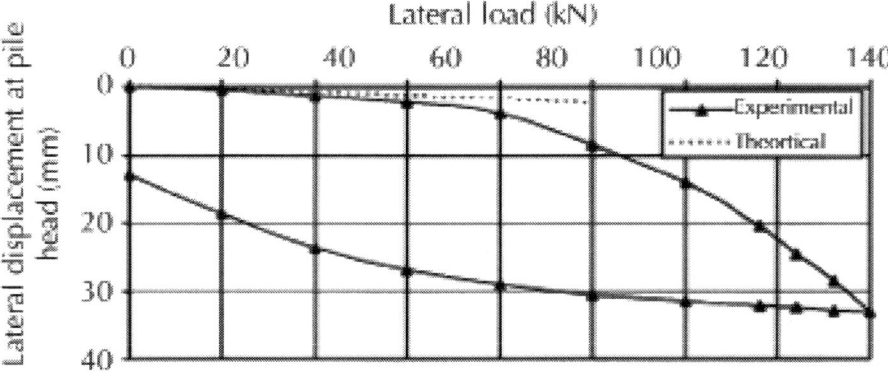

Figure 3: Lateral load versus lateral displacement, test no. 1.

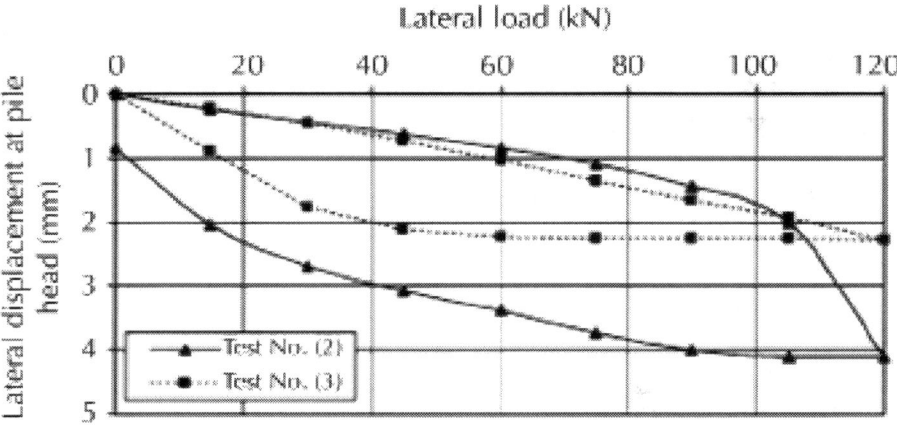

Figure 4: Lateral load versus displacement, test nos. 2 & 3.

In another construction site nearby Alexandria city, Egypt, two field pile loading tests were conducted on two individual piles. Fig. 4 presents the achieved test results. Prior to pile construction, eight boreholes were conducted at the site up to 20 m below the ground surface. Retrieved soil samples from boreholes indicated that the soil consists of a top layer of silty clay with sand up to 9.5 m depth. The top layer overlies sandstone up to 20 m depth, Fig. 5. The shearing parameters of top layer were measured using direct shear box apparatus on undisturbed samples. The obtained parameters are shown in Fig. 5.

Cast in place auger piles of 500 mm diameter and 13 m depth below the ground surface are constructed in the site. The design working lateral load of the pile is 60 kN and the test load 120 kN. The lateral load was applied at the ground surface. Cleary test (2) in Fig. 4 demonstrates that the relationship is linear up to 100 kN. At the same time, test (3) shows that the relationship is almost linear, without any appreciable residual displacement. Taking the case of a vertical pile, the lateral loading on the pile head is initially carried by the soil close to the ground surface. At a low load level, the soil compresses elastically but the lateral movement of the pile is sufficient to transfer some additional incremental loads from the pile to the soil at a greater depth. At a further stage of loading, the soil yields plastically and transfers its load to greater depths. Therefore linear analysis of laterally loaded piles is considered as a good simulation of real behavior of piles under lateral loads within the working load range.

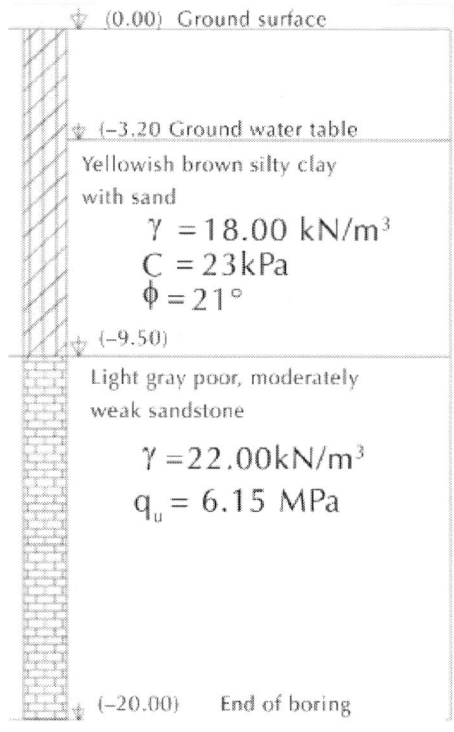

Figure 5: Typical borehole in Alexandria site.

It is worth noting that the tested piles in the two sites were constructed by boring the soil and cast in situ concrete. Therefore, there is a complete contact between the formed pile and the surrounding soil especially near ground surface. The gap that may be formed near ground surface between the pile and the surrounding soil during pile construction as well as the nonlinearity of soil stiffness are the main causes of nonlinearity response of a laterally loaded pile at small values of lateral loads.

The proposed method was also based on that the lateral resistance of a pile in a group is a function of column location belonging to that pile, rather than location within a column of piles, contrary to expectation based on the elastic theory. Rollins et al. [28] and McVay et al. [17] confirmed the above assumptions. Validation of the proposed method is presented in Figure 6a and Figure 6b. The horizontal subgrade reaction was considered to be increased linearly with depth, from zero at the ground surface to ($_h$) at depth 15 m below the ground surface. Selected value of horizontal subgrade reaction was used along with the compiled database to predict the distribution of bending moment along the single individual pile. A comparison between the obtained distribution of bending moment and the measured values by Rollins et al. [28] indicated that the selected value of horizontal subgrade reaction overestimated the maximum bending moment induced in the single pile up to 17%, while LPILE [26] and SWM [2] methods underestimated the induced values up to 20%. Cleary the horizontal subgrade reaction can be used for pile group analysis. The induced bending moment in a pile within a group depends upon the location of the pile in the group.

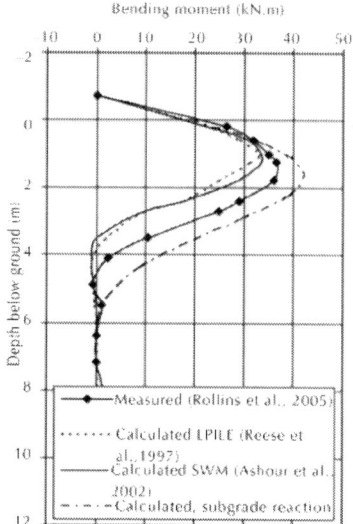

Figure 6a: Comparison between measured and computed bending moments of a single pile, P_H = 24 kN, and $_h$ = 53 MN/m³.

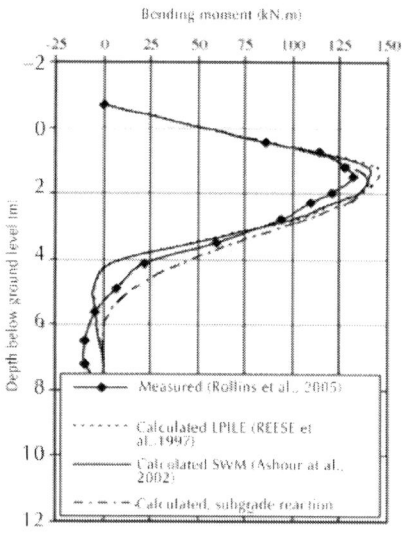

Figure 6b: Comparison between measured and computed bending moments of a single pile, P_H = 89 kN, and $_h$ = 53 MN/m³.

According to the proposed method, the distribution of the bending moment can be obtained by analyzing single individual pile using softening modulus of subgrade reaction that can be obtained by implementing p-multiplier to ($_h$) in order to get ($_{hp}$). The effects of p-multiplier on induced bending moment and the lateral displacement at pile head are shown in Figure 7a and Figure 7b. The applied lateral load at the pile head is 80 kN, while the horizontal subgrade reaction is increasing with depth from zero at the ground surface to a value of 163 MN/m³ at depth 15 m below the ground surface. The pile diameter is 600 mm. As p-multiplier decreased, the soil gets soft and consequently the pile head deflection and the bending moment increased linearly.

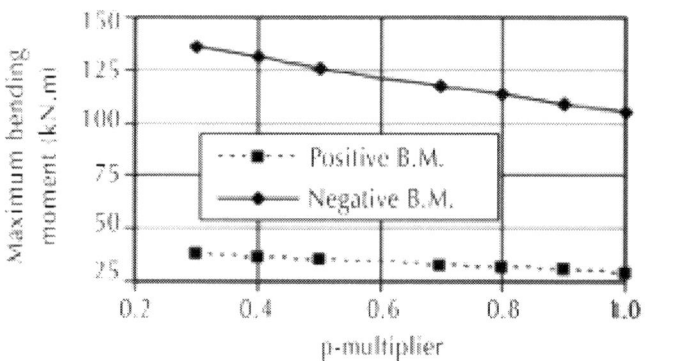

Figure 7a: Bending moments versus p-multiplier.

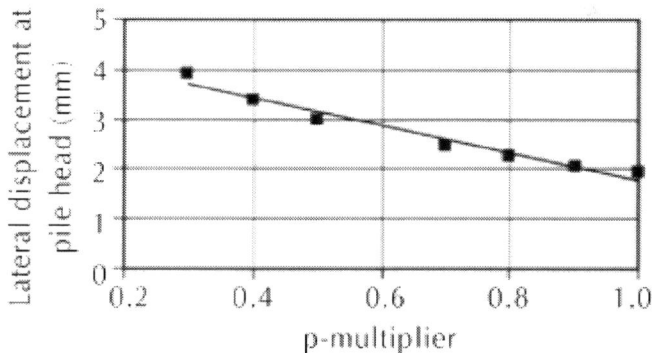

Figure 7b: Lateral displacement at pile head versus p-multiplier.

CONCLUSIONS

This paper presents a simplified method for the analysis of pile groups subjected to lateral loads. The suggested method implements *p*-multiplier factors in combination with the horizontal modulus of subgrade reaction. Interaction effects in closely spaced piles in a group were taken into consideration. The proposed method estimates the distribution of lateral loads among piles in a group and predicts the safe design lateral load of a pile group. The study showed that, laterally loaded piles in sand can be analyzed within the working load range assuming a linear relationship between lateral load and lateral displacement at pile head. The induced maximum bending moments and lateral displacements at pile head of laterally loaded piles decreased linearly as the values of *p*-multiplier increased. Finally, the study indicates that the effective depth of a flexible laterally loaded pile embedded in cohesionless soil is about 16 times the pile diameter.

REFERENCES

1. M. Ashour, G. Norris, Lateral loaded pile response in liquefiable soil, J. Geotech. Geoenviron. Eng. 129 (5) (2003) 404–414.

2. M. Ashour, G. Norris, P. Pilling, Strain wedge model capability of analyzing behavior of lateral loaded isolated piles, drilled shafts, and pile groups, J. Bridge Eng. 7 (4) (2002) 245–254.

3. O. Barton, Laterally Loaded Model Piles in Sand; Centrifuge Tests and Finite Element Analyses, PhD. Thesis, University of Cambridge, 1982.

4. A. Brown, C. Reese, W. O'Neill, Cyclic lateral loading of a largescale pile group, J. Geotech. Eng. 113 (11) (1987) 1326–1343.

5. A. Brown, C. Morrison, C. Reese, Lateral load behavior of pile groups in sand, J. Geotech. Eng. 114 (11) (1988) 1261–1276.

6. B. Broms, The lateral resistance of piles in cohesionless soil, J. Soil Mech. Found. Div. 90 (SM3) (1964) 123–156.

7. F. Castelli, M. Maugeri, Simplified approach for the seismic response of a pile foundations, J. Geotech. Geoenviron. Eng. 135 (10) (2009) 1440–1451.

8. G. Dodagoudar, A. Boominathan, S. Chandrasekaran, Group interaction effects on laterally loaded piles in clay, J. Geotech. Geoenviron. Eng. 136 (4) (2010) 573–582.

9. Federal Highway Administration (FHWA), Design and Construction of Driven Pile Foundation, Publication No. FHWA-HI-97-13, 1996.

10. D. Frechette, K. Walsh, W. Houston, Review of design methods and parameters for laterally loaded groups of drilled shafts, Deep Found. 2002 (2002) 1261–1274.

11. K. Gaaver, Behavior of laterally loaded piles in cohesionless soils, in: The Tenth East Asia–Pacific Conference on Structural Engineering and Construction, 2006.

12. T. Ilyas, F. Leung, K. Chow, S. Budi, Centrifuge model study of laterally loaded pile groups in clay, J. Geotech. Geoenviron. Eng. 130 (3) (2004) 274–283.

13. S. Kumar, L. Lalvani, Lateral load–deflection response of drilled shafts in sand, Int. Eng. J. 84 (2004) 282–286.

14. D.S. Liyanapathirana, H.G. Poulos, Pseudostatic approach for seismic analysis of piles in liquefying soil, J. Geotech. Geoenviron. Eng. 131 (12) (2005) 1480–1487.

15. H. Matlock, L. Reese, Foundation analysis of offshore pile supported structures, in: 5th Int. Conf. on Soil Mech. And Found. Eng., no. 2, Paris, 1961, pp. 91–97.

16. H. Matlock, Correlation for design of laterally-loaded piles on soft clay, in: Proc., 2nd Annual offshore Technology Conf., vol. 1, 1970, pp. 557–594.

17. M. McVay, L. Zhang, T. Molnit, P. Lai, Centrifuge testing of large laterally loaded pile groups in sand, J. Geotech. Geoenviron. Eng. 124 (10) (1998) 1016–1026.

18. L. Mokwa, M. Duncan, Discussion of centrifuge model study of laterally loaded pile groups in clay' by T. Ilyas, F. Leung, K. Chow, & S. Budi, J. Geotech. Geoenviron. Eng. 131 (10) (2005) 1305–1308.

19. K. Ooi, F. Chang, S. Wang, Simplified lateral load analyses of fixed-head piles and pile groups, J. Geotech. Geoenviron. Eng. 130 (11) (2004) 1140–1151.

20.   G. Poulos, Behavior of laterally loaded piles, J. Soil Mech. Found. Div., ASCE 97 (5) (1971) 711–731.

21.   G. Poulos, H. Davis, Pile Foundation Analysis and Design, John Wiley & sons, Inc., New York, NY, 1980.

22.   S. Rao, V. Ramakrisha, M. Rao, Influence of rigidity on laterally loaded pile groups in marine clay, J. Geotech. Geoenviron. Eng. 124 (6) (1998) 542–549.

23.   F. Randolph, The response of flexible piles to lateral loading, J. Geotech. 31 (2) (1981) 247–259.

24.   C. Reese, R. Cox, D. Koop, Analysis of laterally loaded piles in sand, in: Offshore Technology Conference, Houston, 1974, pp. 473–483.

25.   C. Reese, H. Matlock, Non-dimensional solutions for laterally loaded piles with soil modulus assumed proportional to depth, in: 8th Texas Conference on Soil Mech. and Found. Eng., 1956, Austin, pp. 1–41.

26.   C. Reese, T. Wang, A. Arrellaga, J. Hendrix, LPILE Plus 3.0 Windows, Ensoft, Inc., Austin, Tex., USA, 1997.

27.   C. Reese, T. Wang, L. Vasquez, Computer Program GROUP Version 7, Tech. Manual, Ensoft, Inc., Austin, USA, 2006.

28.   M. Rollins, D. Lane, M. Gerber, Measured and computed lateral response of a pile group in sand, J. Geotech. Geoenviron. Eng. 131 (1) (2005) 103–111.

29.   M. Rollins, T. Peterson, J. Weaver, Lateral load behavior of full-scale pile group in clay, J. Geotech. Geoenviron. Eng. 124 (6) (1998) 468–478.

30.   R. Sogge, Laterally loaded pile design, J. Soil Mech. Found. Div., ASCE 107 (9) (1981) 1179–1198.

31.   M. Tomlinson, J. Woodward, Pile Design and Construction Practice, Taylor & Francis Publisher, USA, 2008.

32.   K. Yang, R. Liang, Numerical solution of laterally loaded piles in a two-layer soil profile, J. Geotech. Geoenviron. Eng. 132 (11) (2006) 1436–1443.

Design Considerations for Pile Groups Supporting Marine Structures with Respect to Scour

Yasser E. Mostafa

Department of Irrigation and Hydraulics, Faculty of Engineering, Ain Shams University, Cairo, Egypt

ABSTRACT

Piles supporting marine structures such as jetties, relieving platforms, quay walls and fixed offshore structures are subjected to lateral loads due to berthing and mooring forces, wind, waves, storm surges and current forces. This paper presents some factors that affect the design of pile groups supporting marine structures founded in cohesionless soils.

Some main aspects that should be considered in the pile group design are addressed such as pile batter angle, pile group arrangement, pile spacing, pile slenderness ratio and magnitude of lateral static loading. Numerical analyses were conducted to investigate these design aspects with and without impact of scour. Different scour depths were considered to cover the possible root causes of scour around pile groups such as waves, current and ship propeller jets. The study revealed that scour has greater impact on lateral loading of pile groups compared to its impact on single piles. Pile groups with side-by-side arrangement exposed to scour are more critical than single piles and piles groups with tandem arrangement due to the combined effect of scour and pile-soil-pile interaction. It is also concluded that scour protection is not always required. More attention and considerations should be given to scour protection around piles especially if the piles are closely spaced, arranged side-by-side and if slenderness ratio is less than 12.5.

INTRODUCTION

Marine structures such as jetties, seawalls, relieving platforms, quay walls and fixed offshore (jacket type) structures are often supported on pile groups. The foundation piles usually comprise a large portion of the marine structure cost. These piles are usually subjected to large lateral loads induced from waves, currents, vessel berthing and mooring forces. Also, these piles are subjected to scour due to waves, current and ship propeller jets.

Marine structures and bridge piers supported on pile groups can fail due to severe scour. Numerous publications are found in the literature for investigating the scour around piles for bridge piers and a smaller number of publications investigating the scour around marine structures. Moreover, a very limited number of publications regarding the effect of scour on the behavior of pile groups is found in the literature. Vertical pile capacity is composed of friction along pile length and end bearing at pile toe while pile lateral capacity highly depends on the soil conditions surrounding the top one third of pile length. Therefore, scour impact on lateral pile capacity is more significant than the scour impact on vertical pile capacity.

Global scour refers to a general lowering of the ground surface over a wide area. Figure 1 shows global scour around a fixed offshore structure supported on piles.

Scour around piles varies due to the root cause. For single piles, scour depth in sandy soils (d_s) is 1.3 times pile diameter (d) with a mean of 0.7 [1]. In another meaning, the ultimate scour depth is about 2 times the pile diameter (i.e., $d_s/d = 2$). However, this value can be different for scour due to waves only or due to ship propeller jets only. Some research has been performed to examine the scour around pile groups due to waves and currents such as Sumer and Fredsøe [2], Mostafa and Agamy [3]. However, very limited research has examined how to eva luate the effects of scour on the behavior of pile groups. Among this limited research, recent research focused on piles supporting bridge piers such as Lin et al. [4].

Figure 1: Global scour: wide depression around a jacket structure (after Whitehouse [10]).

Effect of scour on lateral loading of single piles has been investigated in a few recent publications such as Kishore et al. [5], Lin et al. [6], Mostafa [7] and Ni et al. [8]. Mostafa [7] reported that scour has significant impact on single piles installed in sand and less significant impact on piles installed in clay. Also, Mostafa [7] reported that ultimate lateral capacity for single pile subjected to global scour is found to be about 50% to 70% of ultimate lateral capacity of pile subjected to local scour depending on the scour hole dimension.

As piles are usually installed in groups, it becomes necessary to study the effect of scour on the behavior of pile group not just single piles. The combination of scour and pile-soil-pile interaction (i.e., group effect) can lead to a significant reduction in lateral pile capacity and consequently may lead to the failure of marine structures.

This paper presents the impact of global scour around batter and vertical pile groups installed in medium dense sand. The software program GROUP V.7.0 [9] was used in the analyses. For calibration, results from the software were compared with experimental tests on batter pile groups found in the literature.

Different design parameters were investigated in this paper such as pile batter angle, pile group arrangement, pile spacing and pile slenderness ratio. The impact of scour depth on all these parameters was investigated. Only global scour was considered in this study as it has more significant impacts compared to local scour.

Based on the results of the numerical analyses, this paper also provides general recommendations and guidelines on the necessity of using scour protection. Scour protection using riprap or geotextile may be necessary sometimes and may be a waste of money in other cases. The decision to protect the piles from scour depends on the maximum anticipated scour depth based on the root cause of scour and also depends on the pile, soil and loading conditions.

NUMERICAL ANALYSIS AND VALIDATION WITH PREVIOUS EXPERIMENTS

In this paper, numerical analyses were conducted to investigate several parameters affecting the design of pile groups supporting marine structures subjected to scour. The software program Group [9] was used in the analysis. GROUP is a 3D software program for analyzing pile groups subjected to axial and lateral loads. A solution requires iteration to accommodate the nonlinear response of each pile in the group model. The program GROUP solves the nonlinear response of each pile under combined loadings. For closely-spaced piles, the pile-soilpile interaction is taken into account by introducing reduction factors for the p-y curves used for each single pile. These reduction factors or called "p-multipliers" are generated based on results of laboratory and field experiments published in the literature.

A comparison between the results from the computer program Group [9], results from the program Piglet [11] and experimental tests of batter pile groups performed by Zhang et al. [12] was conducted. Batter piles are widely used to support marine structures especially for structures subjected to relatively large lateral loads. Zhang et al. [12] carried out 18 different lateral load tests in the centrifuge on 3 × 3 and 4 × 4 fixed head battered pile groups to investigate the effects of vertical load on the group lateral resistance in cohesionless soils. Zhang et al. [12] proved that designs based on standard lateral load tests with small vertical dead loads would be on the safe side.

Figure 2 shows a sketch of the prototype 3 × 3 battered pile group simulated by the centrifuge models. Two pile arrangements were simulated for the 3 × 3 pile group (Zhang et al. [12]). In the first arrangement, the side pile rows were battered forward at 1:8 slope and the middle row was battered reverse at 1:4 slope (referred to as the 6F3R arrangement). In the second arrangement, the side pile rows were battered reverse at 1:8 slope and the middle row was battered at 1:4 slope (3F6R arrangement). The piles were square aluminum with 304 mm in length and 9.5 mm in width. In prototype scales, the width, total length and embedded length of the piles were 0.43, 13.7 and 10.4 m, respectively and the free length from the ground surface to the point of lateral load application was 2.7 m. The piles were three-diameter spaced and rigidly attached to the pile cap. The soil comprised medium dense sand with relative density (D_r) of 55%.

Figure 3(a) shows the load-deflection curve for the 3 × 3 battered piles with 3F6R arrangement. Results from the computer programs Group and Piglet provide reasonable estimate in comparison with the centrifuge experiments. For low lateral loads, the computer programs tends to slightly overestimate the lateral deflection compared to experimental results; while for the large lateral loads, the computer programs tend to underestimate the lateral deflections. Figure 3(b) indicates that, for the same lateral load, pile group with 6F3R arrangement has less deflection than that for the 3F6R arrangement. This can be attributed to the fact that in the case of 6F3R arrangement six piles have steeper batter angle slope (1:4) and three piles have milder batter angle slope (1:8) which provides higher lateral resistance than the case of 3F6R arrangement.

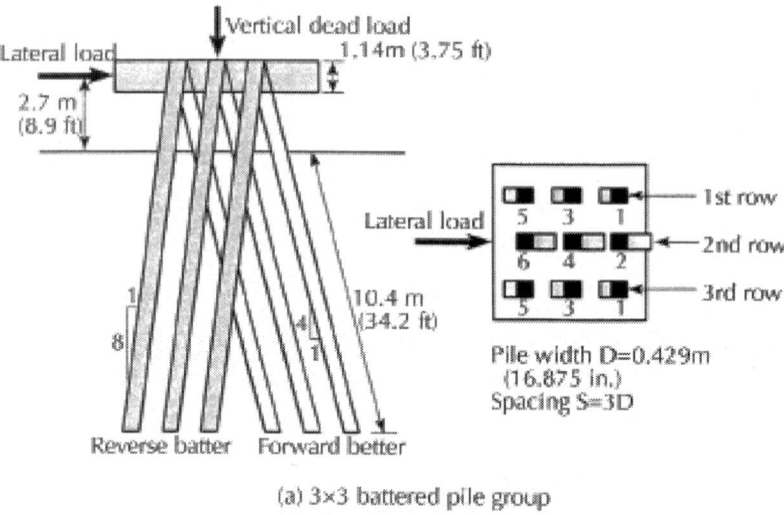

(a) 3×3 battered pile group

Figure 2: Prototype 3 × 3 battered pile groups tests (after Zhang et al. [12]).

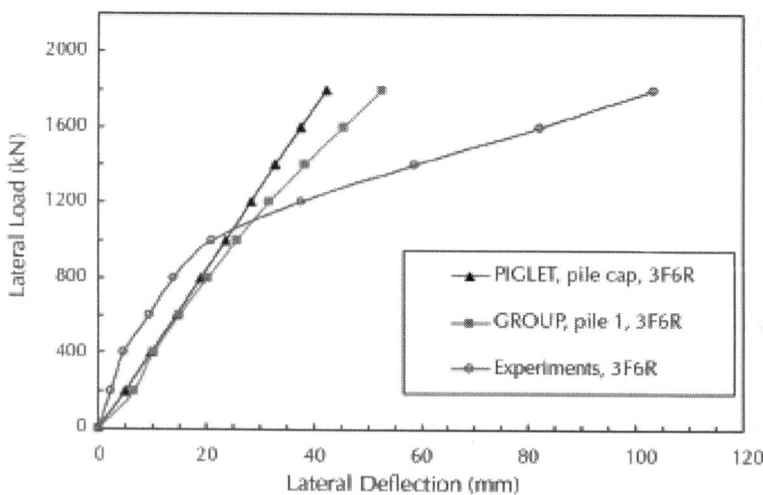

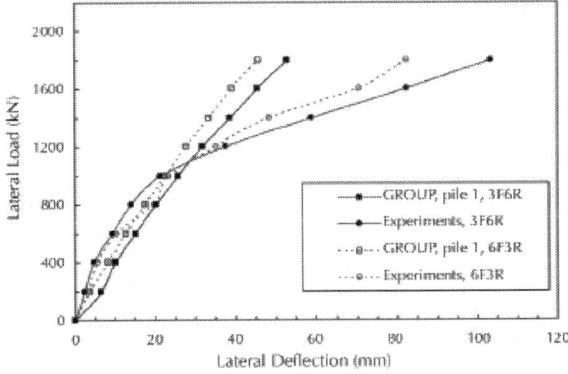

Figure 3: Load-deflection curves for the 3 × 3 battered pile groups considering two pile arrangements: (a) 3F6R arrangement and (b) 6F3R arrangement.

EFFECT OF BATTER ANGLE

Batter (raked) piles are widely used to support marine structures as they are effective in restricting horizontal displacement of the foundation subjected to large lateral loads or in a liquefied soil flow environment. The disadvantages of such piles are the larger axial forces in the piles [13]. The forces on piles supporting coastal structures are axial vertical loads due to own weight, loads from trucks and cranes and lateral loads from ship impacts, wave loads and mooring forces. The use of batter piles along with vertical piles increases the overall pile group efficiency. As reported by Rajashree and Sitharam [14], Batter piles are classified as positive batter (slip surface deflects upward) and negative batter (slip surface deflects downward) depending on the formation of slip surfaces [15].Figure 4 illustrates the negative and positive batter piles.

In this paper, the effect of batter angle was investigated. In the analysis, all piles were assumed to be steel piles with outside diameter (d) of 0.5 m, total pile length (L) of 30 m and cross section area (A) of 0.01944 m^2. The pile spacing to diameter ratio (S/d) was taken to be 3. The distance between point of lateral load at the jetty deck and the seabed (i.e., pile free length, L_f) was assumed to be 5 m. This free length is required for vessels berthing and it varies depending on the vessel draft, under keel clearance and distance between sea level and jetty deck level. All pile heads were assumed to be fixed to the jetty deck. The soil was assumed to be medium dense sand with a unit weight (g) of 17 kN/m³ and angle of internal friction (f) of 35°. The ratio between pile and soil modulus of elasticity (E_p/E_s) was assumed to be 2000. In the analysis, two piles were considered; one pile has negative batter and the other one has positive batter. Different batter angles of 0°, 5°, 10° and 20° were investigated.

Load-displacement curves for different pile batter angles are shown in Figure 5. For the same lateral loading on pile group, a small batter angle significantly reduces the lateral pile displacement. For example, for a lateral load of 400 kN applied to the group at the deck level, the normalized lateral pile displacement (y/d) is 0.23 for vertical piles (b = 0°) and about 0.1 for batter piles with angle b of 5° (i.e., reduction of 57% in pile head displacement). It is also noted that increasing the batter angle from b = 0° to b = 5° increases the lateral capacity from 400 kN to 700 kN (i.e., increase of 75% of lateral capacity).

Figure 6 shows the lateral displacement and bending moment along pile length if the pile group is subjected to lateral load of 100 kN applied at the pile cap. A slight increase in the batter angle significantly reduces the lateral pile displacement and bending moment. It is worth mentioning that bending moment and deflection of positive batter piles are almost similar or slightly higher than that for negative batter piles. This may be attributed to the pile head conditions which were assumed to be fixed in all analyses. Due to scale limitations, only negative piles are presented.

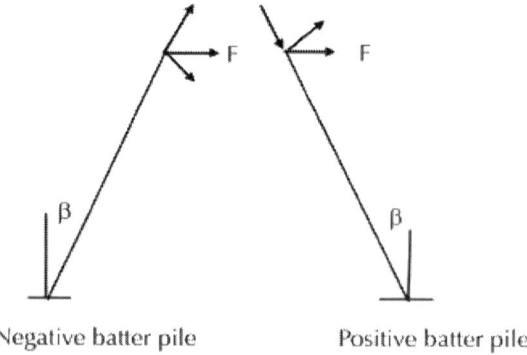

Negative batter pile   Positive batter pile

Figure 4: Sketch of negative and positive batter piles.

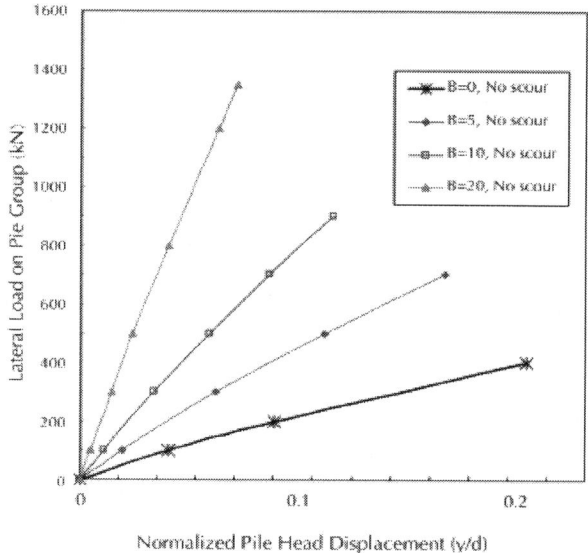

Figure 5: Load-displacement curves for different pile batter angle.

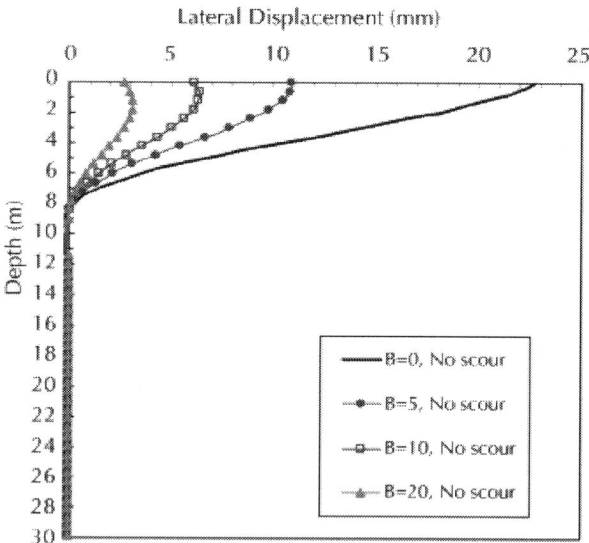

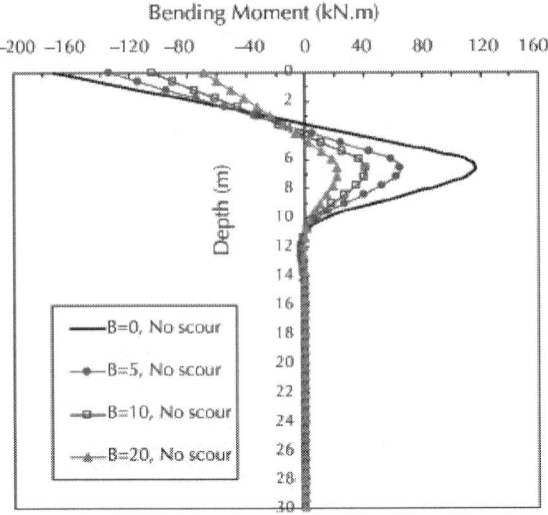

Figure 6: Effect of batter angle on the pile response (F = 100 kN) (a) Shear force along pile length; (b) Bending moment along pile length.

Figure 7 shows that lateral pile group capacity significantly increases by the increase in batter angle. A batter angle of 5° increases the lateral capacity by about 40% over vertical piles (b = 0). The increase in the group capacity is almost linear. The rate of increase in capacity if b ranges between 10° and 20° is higher than that rate if b ranges between 0° and 10°. If scour depth equivalent to twice pile diameter ($d_s/d = 2$) is considered, the group lateral capacity decreases by about 5% to 10%. The rate of change decreases with the increase in batter angle. In another meaning, the scour becomes less significant with the increase in batter angle.

Wave loading is cyclic in nature. In offshore conditions the number of cycles of wave loading may be up to 600 [14]. The cyclic loading causes a gap between the pile and soil interface. Effect of cyclic loading on the behavior of pile groups was examined in this paper. No major difference was found in the results for the cyclic loading conditions compared to static loading conditions. This is because the analysis was performed in sandy soils as sand tends to fill the gap between the pile and soil interface during cyclic loading. For cohesive soils, cyclic loading is expected to have larger impact on the piles compared to static loading.

Caution should be exercised for seismic loading conditions as it has been reported in some publications that batter piles have poor performance during recent earthquakes [16].

EFFECT OF SCOUR ON VERTICAL PILE GROUPS

Effect of global scour on vertical pile groups was investtigated. Several parameters were considered including the variation in scour depth. The effect of scour depth combined with other parameters such as different pile arrangement, spacing between piles and pile slenderness ratio were investigated.

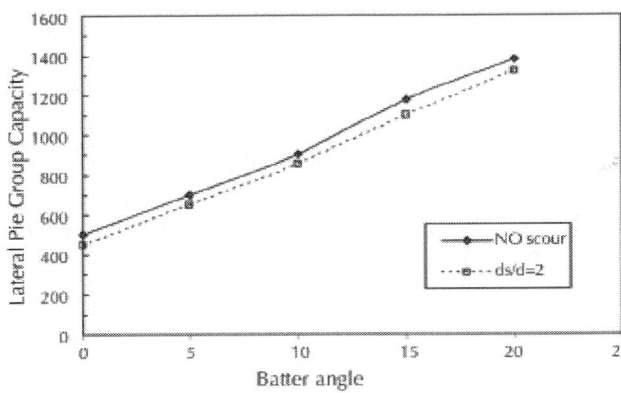

Figure 7: Relationship between pile batter angle and the lateral pile group capacity (d = 0.5 m, L = 30 m, L_f = 5 m, S/d = 3).

The same soil and pile characteristics described in Section 3 were assumed except that all piles are considered to be vertical. All pile groups comprised two piles. No free length was assumed in the analysis of vertical piles.

Assuming the piles are Grade 4 steel with a yield stress of 450 MPa, the equivalent moment causing the pile to yield is about 1040 kN/m. The ultimate lateral pile capacity corresponds to the load that causes the yield moment.

Scour Depth around Pile Groups

The scour depth around pile groups is based on the root cause of scour. Piles supporting coastal and marine structures are subjected to scour due to waves, currents and ship propeller jets. A summary of the expected scour depth due to these causes is provided in the following subsections.

Scour Due to Waves and Current

Some experimental research has been performed to examine the scour around pile groups due to waves and currents. Sumer and Fredsøe [2] examined the wave scour around a group of closed spaced piles. Sumer et al. [17] investigated the global and local scour at pile groups exposed to steady currents. Mostafa and Agamy [3] examined the combined effect of waves and current on a group of closed and widely spaced piles. Scour depth to pile diameter ratio (d_s/d) due to waves only can be as high as 1 for Keulegan-Carpenter number (KC) of 13. For pile groups exposed to currents only, total scour (global scour and local scour) increases with the increase in number of piles. The scour depth to pile diameter ratio (d_s/d) can be as high as 2.6 [17].

Scour Due to Ship Propeller Jets

During berthing or de-berthing operations, a ship is in proximity to harbor structures and erosion may occur around piles supporting jetties or relieving platforms and sloping riprap [18]. Figure 8 shows a sketch of scour around piles supporting jetties or relieving platforms due to ship propeller jets. It is noticed that scour at the front (leading) pile is larger than that at the trailing (aft) pile. If pile spacing increases, scour at the trailing pile decreases as a result of the deposition of scoured sediment taking place at the vicinity of the front pile as can be seen in Figure 8.

Yuksel et al. [18] concluded that maximum scour depth and sand deposition for pile groups vary significantly with densimetric Froude number Fr_d, pile spacing to diameter ratio (S/d) and pile diameter to jet diameter ratio (d/d_o). Scour at the front (leading) pile is larger than that at the trailing (aft) pile. Empirical equations for scour calculations due to ship propeller jets can be found in the literature. Experimental results of Yuksel et al. [18] indicated that maximum scour depth to pile diameter ratio (d_s/d) is about 2.25.

In this paper, normalized scour depth (d_s/d) between 0 and 3 was considered in the analysis to cover the different root causes of scour around piles supporting coastal structures.

Effect of Pile Arrangement

The analysis considered two main pile groups. The pile groups were assumed to be arranged side-by-side (i.e., q = 90°) or tandem arrangement (i.e., q = 0°) where q is the angle between the lateral load and the line connecting the two piles and S is the centre line to centre line spacing between piles. Figure 9 shows a general sketch indicating a pile location within a pile group subjected to lateral loading. The piles were assumed to be spaced three times the pile diameter (i.e., S/d = 3). Pile-soil-pile interaction was considered in the analysis. The reduction factors generated in the program Group [9] were used in this study. The length to diameter ratio (L/d) was assumed to be 60 so that pile embedment length has no impact on the results.

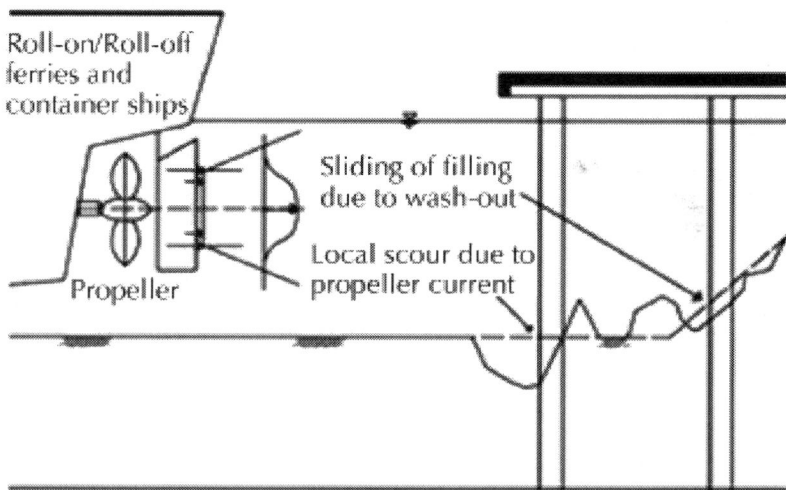

Figure 8: Erosion of seabed due to ship propeller jets (Chin et al., 1996 [19]).

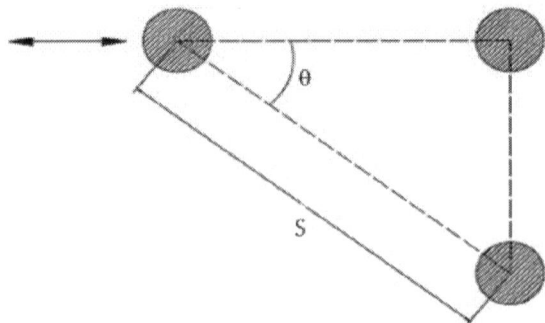

Figure 9: The location of a pile in pile group under lateral loading.

Figure 10 shows the effect of normalized scour depth (d_s/d) on the lateral capacity of a group of two piles with side-by-side arrangement and tandem arrangement. Figure 10(b) indicates that the reduction in pile group capacity of groups with side-by-side arrangement is higher than that for groups with tandem arrangement. A relatively small scour depth ($d_s/d = 1$) leads to a reduction in lateral group capacity ranging between about 12.5% and 15% for piles with tandem arrangement and side-by-side arrangement, respectively. A scour depth of three pile diameter (i.e., $d_s/d = 3$) causes a reduction in the lateral group capacity ranging between 29% and 35% for piles with tandem arrangement and side-by-side arrangement, respectively. The rate of reduction in lateral group capacity if d_s/d varies between 0 and 1 is slightly higher than that if d_s/d is between 1 and 3.

Figure 11 shows load-displacement curves for groups of two piles with tandem arrangement ($q = 0°$) and side-by-side arrangement ($q = 90°$) neglecting and considering scour. Load-displacement curves for single pile computed using program LPILE were plotted for comparison. It can be seen that the load-displacement curve for single pile is very similar to that for two piles with side-by-side arrangement. This is because the side-by-side reduction factor due to group interaction is negligible if the spacing is greater than three times the pile diameter [Group] [9]. For the same lateral loading, pile group with tandem arrangement has less displacement than that for pile group with side-by-side arrangement. For relatively small lateral loads on the group (F < 100 kN) no major differences in the normalized pile head displacement (y/d) are found due to pile group arrangement. The impact of pile arrangement is more pronounced with the increase in lateral loading due to the nonlinearity in the pile-soil system. For single pile and pile groups with different arrangements, scour increases the pile head displacement. For the same lateral loading, piles with tandem arrangement even when scour is considered has lower displacement than the piles with side-by-side arrangement with no scour. This is due to the greater reduction in group efficiency for side-by-side arrangement compared to the reduction in group efficiency for tandem arrangement.

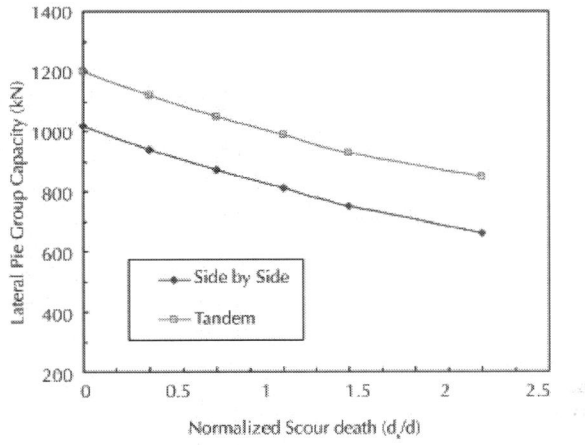

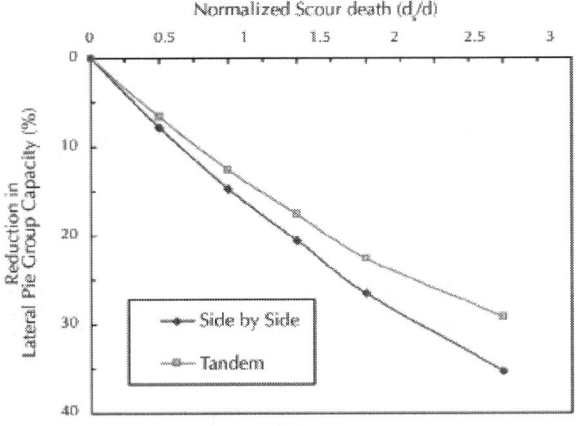

Figure 10: Relationship between normalized scour depth and lateral pile capacity (S/d = 3, L = 30 m, L/d = 60).

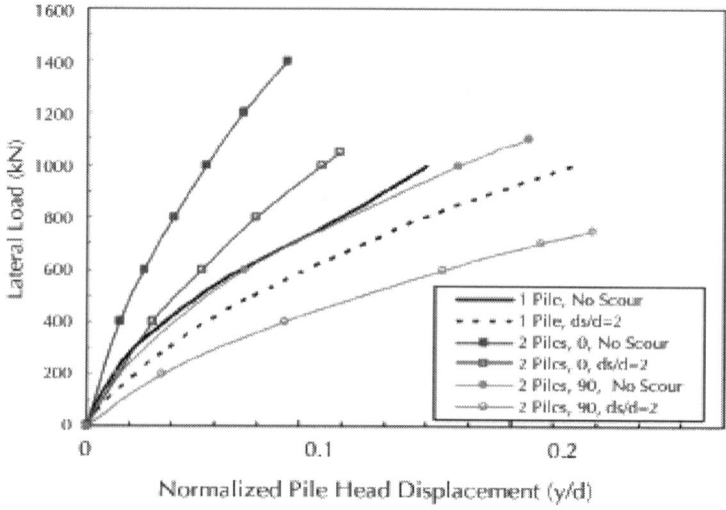

Figure 11: Load-displacement curves for single pile and pile groups with different arrangement neglecting and considering scour (S/d = 3, L = 30 m, L/d = 60).

Figure 12 shows lateral displacement and bending moment along pile length when a group of two piles is subjected to lateral load (F) of 200 kN. For the sake of comparison, single pile subjected to lateral load of 100 kN was computed and plotted. It is noted that displacement of piles with side-by-side arrangement (q = 90°) is higher than that for single pile and pile group with tandem arrangement (q = 0°). Scour depth of twice pile diameter increases the head displacement of pile group with side-by-side arrangement from 7.4 mm to 18 mm (i.e., increase by about 114%). For piles with tandem arrangement, the displacement at pile head increases due to scour from about 3.4 mm to 6.5 mm (i.e., increase by about 91%). For the single pile, scour increased the pile head displacement from about 2.6 mm to about 4.9 mm (i.e., increase by about 88%). Under the same loading conditions, the displacement of pile group is higher than that for single pile. This is attributed to the pile-soil-pile interaction. The scour effect on pile group with side-by-side arrangement is more significant than that for single piles and pile group with tandem arrangement. The bending moment for piles with side-by-side arrangement is larger than that for single pile and pile group with tandem arrangement.

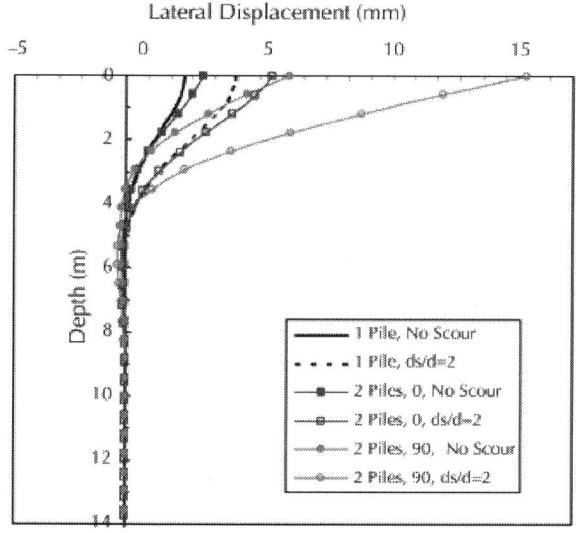

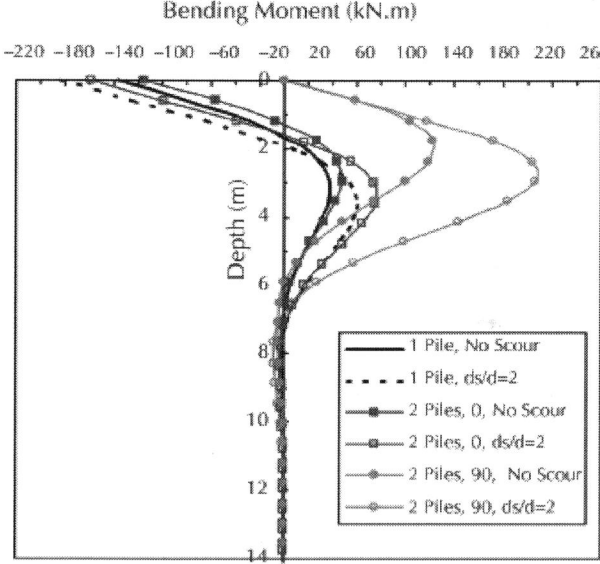

Figure 12: Effect of pile arrangement on lateral response along pile length, (a) Lateral displacement; (b) Bending moment (F = 200 kN, S/d = 3, L = 30 m, L/d = 60).

Effect of Pile Spacing

The effect of centre line to centre line pile spacing considering pile-soil-pile interaction was evaluated. Spacing to diameter ratio (S/d) of 1, 3 and 5 for pile groups with tandem and side-by-side arrangements was considered. The impact of scour is also investigated.

When piles are closely spaced, the shear failure planes resulting from the movement of each pile will overlap and the ultimate resistance for a pile in a group may be less than that of a single pile. This is called "shadow effect" or "pile-soil-pile interaction" which influences the efficiency of individual piles in a group. Reduction factors generated in the program Group [9] were used in this study. These reduction factors are based on many laboratory and field experiments collected from the literature.

Brown et al. [20] introduced the concept of the pmultiplier. This concept represents the response of the group to lateral loading in terms of the response of an assembly of single piles with the soil reaction modeled using p-y curves adjusted using a "p-multiplier" where p is the lateral soil reaction and y is lateral pile displacement. The p-multiplier assumes a different value that depends on whether a pile is in a leading or in a trailing position and the angle between the line connecting two piles and the load direction.

Figure 13 shows the relationship between normalized scour depth and lateral pile group capacity taking into consideration the reduction in group efficiency due to pile-soil-pile interaction. It is evident that the lateral pile group capacity for the two piles with tandem arrangement is higher than that of the two piles with side-by-side arrangement by about 18% to 32%. For closely spaced piles (i.e., S/d = 1) the group capacity is reduced by about 11% and 5% for case of side-by-side arrangement and tandem arrangement respectively. A scour depth equivalent to three times pile diameter decreases the lateral group capacity by about 32% to 35% for groups with side-by side arrangement. The same scour depth reduced the capacity by about 27% to 30% for groups with tandem arrangement.

It should be noted that for tandem arrangement, the ultimate lateral capacity of trailing pile is lower than that of leading pile assuming the scour depth is the same at leading and trailing piles. Therefore, lateral load that causes yield moment of trailing pile was considered in Figure

13.

Figure 14 shows load-displacement curves for two pile group with side-by-side arrangement considering different pile spacings. Impact of scour was also considered by assuming a scour depth equivalent to twice the pile diameter. Figure 14 indicates that pile displacement increases significantly when the piles are closely spaced and scour is considered. As expected, the ultimate group capacity increases with the increase in pile spacing and if scour is neglected. The effect of scour is more pronounced with the increase in lateral loading. As an example, a lateral load of 600 kN on a group with S/d = 1 causes a displacement (y/d) of 0.11 and 0.23 if scour is neglected and considered, respectively. For the same lateral load applied on a group of widely spaced piles (S/d = 5), y/d is computed to be 0.074 and 0.17 if scour is neglected and considered, respectively. For the same pile spacing ratio, the displacement increases by about 100% to 130% due to scour. If scour is neglected and pile spacing ratio is varied, the displacement increases by about 7% if S/d is reduced from 5 to 3 and the displacement increases by about 40% if S/d is reduced from 3 to 1. This is due to the pile-soil-pile interaction. If scour is neglected, the ultimate lateral group capacity is reduced from 1050 kN if S/d = 5 to 900 kN if S/d = 1. This corresponds to a reduction of about 14% due to pile-soil-pile interaction. For the same S/d ratio, the ultimate lateral capacity reduces by about 23% to 29% due to the scour effect.

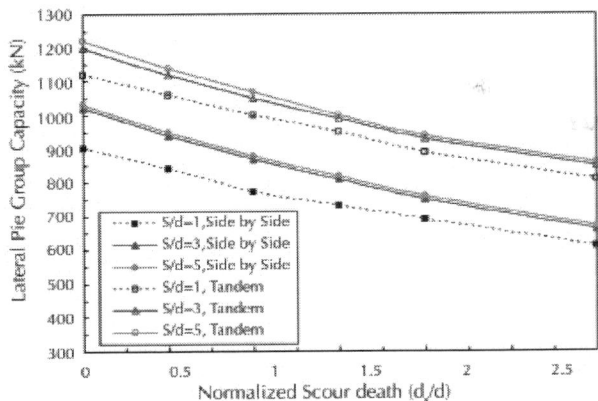

Figure 13: Normalized scour depth versus lateral pile group capacity for different pile spacing and arrangement.

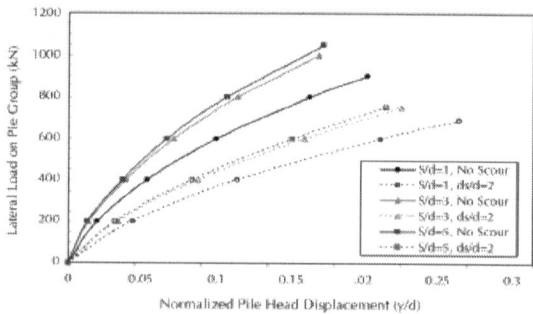

Figure 14: Load-displacement curves for pile group with different pile spacing with and without scour (side-by-side arrangement, L/d = 60).

Therefore, it is concluded that the effect of scour on lateral group capacity and displacement is more pronounced than the effect of pile-soil-pile interaction.

Effect of Pile Slenderness Ratio

Pile length to diameter ratio (slenderness ratio) may be critical for piles supporting coastal structures under scoured conditions. To examine the effect of pile slenderness ratio, L/d ratios of 5, 7.5, 10, 12.5, 15, 20 and 40 were considered. Pile slenderness ratio for some dolphins or suction piles may be as low as 5. The pile spacing ratio (S/d) was kept constant as 3. Groups of two piles with side-by-side arrangement were considered as this is a more critical case than groups with tandem arrangement.

Figure 15 shows the relationship between normalized scour depth and lateral pile capacity for different pile slenderness ratio. It is noted that for L/d = 5, the rate of decrease in lateral pile group capacity between d_s/d = 0 to 0.5 is higher than that rate if d_s/d varies between 0.5 and 3. For L/d = 7.5 and 10, the rate of decrease in lateral group capacity between d_s/d = 0 and 1 is higher than that rate if d_s/d varies between 1 to 3.

It is also noted from Figure 15 that percentage decrease in lateral pile group capacity (PDC) remains almost constant after the pile slenderness ratio is greater than about 12.5. In another meaning, for pile groups exposed to scour, increasing L/d ratio over 12.5 will not add much difference to the group capacity. Accordingly, the decrease in group capacity due to scouring will be more serious for short piles. This is somewhat similar to a conclusion drawn by Ni et al. [8] who concluded that for single piles, PDC values remain almost constant after L/d is greater than 10.

SCOUR PROTECTION

As discussed in this paper, pile groups subjected to scour due to waves, current and ship propeller jets may be exposed to a scour depth up to 2.6 pile diameter. A value of scour depth to diameter ratio (d_s/d) of 2 is often used in practice. It is important for designers and practitioners to know when scour protection around piles supporting coastal structures may be necessary and when it is not.

According to Mostafa and Agamy [3] and Abdeldayem et al. [21], pile groups with side-by-side arrangement causes more scour than groups with tandem arrangement. Scour depth for some cases of pile groups with side-by-side arrangement increases as much as about two times more than its magnitude for the case of single pile according to Mostafa and Agamy [3] and Sumer et al. [17]. Accordingly, for a jetty or berthing structure supported on piles exposed to scour due to ship propeller jets, the leading pile row, which is arranged side-by-side, is expected to suffer more from scour.

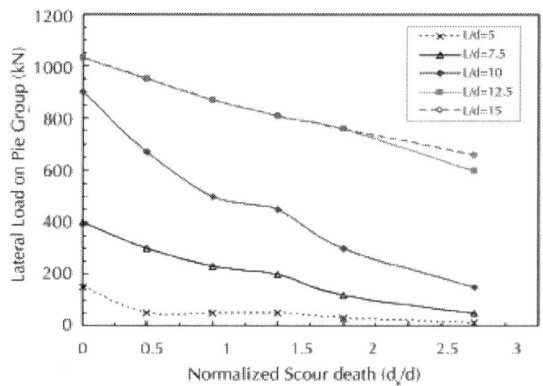

Figure 15: Lateral pile group capacity versus normalized scour depth (ds/d) considering different pile slenderness ratio (L/d) (S/d = 3, side-by-side arrangement).

From a geotechnical prospective, the ultimate lateral capacity for piles arranged side-by-side is lower than that for piles with tandem arrangement. The side-by-side arrangement is exposed to higher lateral deflections and bending moments. Therefore, the combined effect of scour and pile-soil-pile interaction for closely spaced piles makes the side-by-side arrangement the worst case scenario in terms of significant reduction in pile capacity. This scenario may be even worse if the loading arm or free length (distance between jetty deck level and seabed) is high. In this case, scour protection and frequent monitoring of erosion around the piles may be required.

For raked piles, as scour is less significant when the batter pile angle increases, scour protection is not necessary depending on the required lateral pile capacity.

For short piles, consideration should be given to using scour protection measures such as riprap or geotextile. Alternatively, consideration should be given to increasing the pile length so that the slenderness ratio is about 12.5. For already constructed piles with slenderness ratio less than 12.5, consideration should be given to using a suitable scour protection measure.

It should be noted that this section provides general recommendations on when to use scour protection. The decision to use scour protection around piles should be taken based on the pile, soil and loading conditions.

CONCLUSIONS

Some design aspects of pile groups supporting marine structures founded on cohesionless soils were studied using numerical modeling. These aspects are scour depth, lateral static loading, batter pile angle, pile group arrangement, pile spacing and pile slenderness ratio. For the input data assumed in the analyses, the following conclusions can be drawn:

1. A small increase in batter pile angle increases the lateral pile capacity significantly. A batter angle of 5° increases the lateral capacity by about 40% over vertical piles.
2. Scour depth becomes less significant with the increase in pile batter angle.

3. Pile groups with side-by-side arrangement experience higher displacement and bending moments compared to single piles and pile groups with tandem arrangement due to the combined effect of scour depth and pilesoil-pile interaction.

4. The ultimate lateral capacity for pile groups with side-by-side arrangement is lower than that for pile groups with tandem arrangement. For spacing to diameter ratio of 3, a scour depth of three pile diameter (i.e., $d_s/d = 3$) causes a reduction in the lateral group capacity ranging between 29% and 35% for piles with tandem arrangement and side-by-side arrangement, respectively.

5. The rate of reduction in lateral group capacity if d_s/d varies between 0 and 1 is slightly higher than that if d_s/d is between 1 and 3.

6. For relatively small lateral loads on the group (F < 100 kN), no major differences in the normalized pile head displacement (y/d) are found due to the pile group arrangement. The impact of pile arrangement is more pronounced with the increase in lateral loading due to nonlinearity in the pile-soil system.

7. For the same lateral loading, piles with tandem arrangement even when scour is considered has lower displacement than piles with side-by-side arrangement with no scour.

8. The bending moment for piles with side-by-side arrangement is larger than that for single pile and pile group with tandem arrangement.

9. The effect of scour on lateral group capacity and displacement is more pronounced than the effect of pilesoil-pile interaction.

10. The percentage decrease in lateral pile group capacity (PDC) remains almost constant after the pile slenderness ratio is greater than about 12.5.

11. More attention and consideration should be given to scour protection around piles especially if the piles are closely spaced, arranged side-by-side and if slenderness ratio is less than 12.5.

REFERENCES

1. B. M. Sumer, J. Fredsøe and N. Christiansen, "Scour around a Vertical Pile in Waves," Journal of Waterway, Port, Coastal and

Ocean Engineering, ASCE, Vol. 118, No. 1, 1992, pp. 15-31. doi:10.1061/(ASCE)0733-950X(1992)118:1(15)

2.   B. M. Sumer and J. Fredsøe, "Wave Scour around Group of Vertical Piles," Journal of Waterway, Port, Coastal and Ocean Engineering, ASCE, Vol. 124, No. 5, 1998, pp. 248-255. doi:10.1061/(ASCE)0733-950X(1998)124:5(248)

3.   Y. E. Mostafa and A. F. Agamy, "Scour around Single Pile and Pile Groups Subjected to Waves and Currents," International Journal of Engineering Science and Technology, IJEST, Vol. 3, No. 11, 2011, pp. 8160-8178.

4.   C. Lin, C. Bennett, J. Han and R. L. Parsons, "Integrated Analysis of the Performance of Pile-Supported Bridges under Scoured Conditions," Engineering Structures, Vol. 36, 2012, pp. 27-38. doi:10.1016/j.engstruct.2011.11.015

5.   Y. N. Kishore, S. N. Rao and J. S. Mani, "The Behaviour of Laterally Loaded Piles Subjected to Scour in Marine Environment," KSCE Journal of Civil Engineering, Vol. 13, No. 6, 2009, pp. 403-406. doi:10.1007/s12205-009-0403-2

6.   C. Lin, C. Bennett, J. Han and R. L. Parsons, "Scour Effects on the Response of Laterally Loaded Piles Considering Stress History of Sand," Computers and Geotechnics, Vol. 37, No. 7-8, 2010, pp. 1008-1014.

7.   Y. E. Mostafa, "Effect of Local and Global Scour on Lateral Response of Single Piles in Different Soil Conditions," Engineering, Vol. 4, No. 6, 2012, pp. 297-306.

8.   S. Ni, Y. Huang and L. Lo, "Numerical Investigation of the Scouring Effect on the Lateral Response of Piles in Sand," Journal of Performance of Constructed Facilities, Vol. 263, 2012, pp. 320-325. doi:10.1061/(ASCE)CF.1943-5509.0000224

9.   Ensoft Inc., "GROUP, A Program for Analyzing a Group of Piles Subjected to Axial and Lateral Loading," Version 7.0, Technical Manual, Austin, 2006.

10.   R. Whitehouse, "Scour at Marine Structures, A Manual for Practical Application," Thomas Telford Publications, London, 1998. doi:10.1680/sams.26551

11. M. F. Randolph, "PIGLET. A Program for Analysis and Design of Pile Groups," Version 5.1, Technical Manual, The University of Western Australia, Perth, 2004.

12. L. M. Zhang, M. C. McVay, S. J. Han, P. W. Lai and R. Gardner, "Effects of Dead Loads on the Lateral Response of Battered Pile Groups," Canadian Geotechnical Journal, Vol. 39, No. 3, 2002, pp. 561-575. doi:10.1139/t02-008

13. T. Tazoh1, M. Sato, J. Jang and G. Gazetas, "Centrifuge Tests on Pile Foundation-Structure Systems Affected by Liquefaction-Induced Soil Flow after Quay Wall Failure," The 14th World Conference on Earthquake Engineering, Beijing, 12-17 October 2008.

14. S. S. Rajashree and T. G. Sitharam, "Nonlinear FiniteElement Modeling of Batter Piles under Lateral Load," Journal of Geotechnical and Geoenvironmental Engineering, Vol. 127, No. 7, 2001, pp. 604-612. doi:10.1061/(ASCE)1090-0241(2001)127:7(604)

15. G. P. Tschebotarioof, "The Resistance to Lateral Loading of Single Piles and Pile Groups," Special Publication No. 154, ASTM, 1953, pp. 38-48

16. R. E. Harn, "Displacement Design of Marine Structures on Batter Piles," 13th World Conference on Earthquake Engineering, Vancouver, 1-6 Augsut 2004, Paper No. 543.

17. B. M. Sumer, K. Bundgaard and J. Fredsøe, "Global and Local Scour at Pile Groups," Proceedings of the Fifteenth International Offshore and Polar Engineering Conference, Seoul, 19-24 June 2005, pp. 577-583.

18. A. Yuksel, Y. Celikoglu, E. Cevik and Y. Yuksel, "Jet Scour around Vertical Piles and Pile Groups," Ocean Engineering, Vol. 32, No. 3-4, 2005, pp. 349-362.doi:10.1016/j.oceaneng.2004.08.002

19. C. Chin, Y. Chiew, S. Y. Lim and F. H. Lim, "Jet Scour around Vertical Pile," Journal of Waterway, Port, Coastal and Ocean Engineering, Vol. 122, No. 2, 1996, pp. 59-67.doi:10.1061/(ASCE)0733-950X(1996)122:2(59)

20. D. A. Brown, C. Morrison and L. C. Reese, "Lateral Load Behavior of Pile Group in Sand," Journal of Geotechnical Engineering,

ASCE, Vol. 114, No. 11, 1988, pp. 1326- 1343.doi:10.1061/ (ASCE)0733-9410(1988)114:11(1261)

21. A. Abdeldayem, G. Elsaeed and A. Ghareeb, "Effect of Pile Group Arrangements on Local Scour Using Numerical Models," Advances in Natural and Applied Sciences, Vol. 52, 2011, pp. 141-146.

Chapter 9

Mechanical and Acoustic Properties of Magnesium Alloys Based (Nano) Composites Prepared by Powder Metallurgical Routs

Zuzanka Trojanová[1], Pavel Lukáč[1], Zoltán Száraz[2], and Zdeněk Drozd[1]

Charles University in Prague, Czech Republic
The University of Manchester, UK

INTRODUCTION

Magnesium materials are excellent candidates for structural applications where low weight plays an important role. However, Mg and its alloys possess low stiffness and strength compared with aluminum alloys. One way to enhance the strength of Mg is to reinforce it with stronger particles or fibers of the second phase, essentially forming composite microstructures. Composites reinforced with micro-sized particulates of various materials were very often used to enhance the elastic modulus and mechanical properties. Addition of micro-particles, such as SiC, Y_2O_3, MgO, Al_2O_3 particles and CNT (carbon nanotubes) into Mg and Mg alloys has been shown to improve yield strength, modulus, hardness, fatigue and wear resistance, as well as damping properties and thermal stability (Száraz et.al., 2007, Moll et.al., 2000, Ferkel & Mordike, 2001, Trojanová et al., 1997). By scaling the particle size down to the nanometer scale, it has been shown that novel material properties can be obtained (Thostenston et al., 2005).

MATERIALS PROCESSING

Preparation of Mg Alloys Reinforced With SiC Particles

The AZ91, WE54 and Mg-8Li magnesium alloys reinforced with SiC particles with were processed by a powder metallurgy method. Mixing of the matrix alloys powders with SiC microparticles (mp) was carried out first in an asymmetrically moved mixer with subsequent milling in a ball mill. The powder was capsulated in magnesium containers and extruded at 400 °C using a 400 t horizontal extrusion press. The composite samples were not thermally treated. Microstructure of the as prepared WE54/SiC composite in introduced in Fig. 1a (the micrograph was taken perpendicular to the extrusion direction). SiC mp are non-uniformly distributed in the matrix; they form in many cases small clusters. The size of sharp bounded more or less uniaxial particles was approximately 9 µm and the grain size in the matrix about 3-4 µm. Light micrographs and transmission electron micrographs showed no pores in the composite and the binding between SiC particles and the matrix was perfect. No defects were found in the vicinity of SiC particles and no chemical reaction at the interface matrix/SiC particles was observed. Microstructure of the Mg8Li/SiC composite exhibits a mixture of two phases (hexagonal close packed phase and body centred cubic phase). In Fig. 2 light and darker phase are visible together with SiC particles The X-ray analysis revealed the relation between both phases as : = 55 : 45. As the grain size the mean value of 5±2 μm was taken from both phases. Resulting materials contained different volume fraction of particles: AZ91 (13 vol.%), WE53 (13 vol.%) and Mg8Li (7% vol.%).

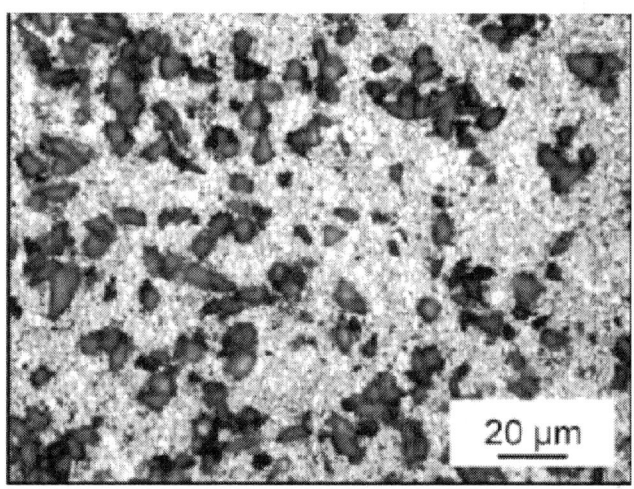

Figure 1: Microstructure of WE54/SiC.

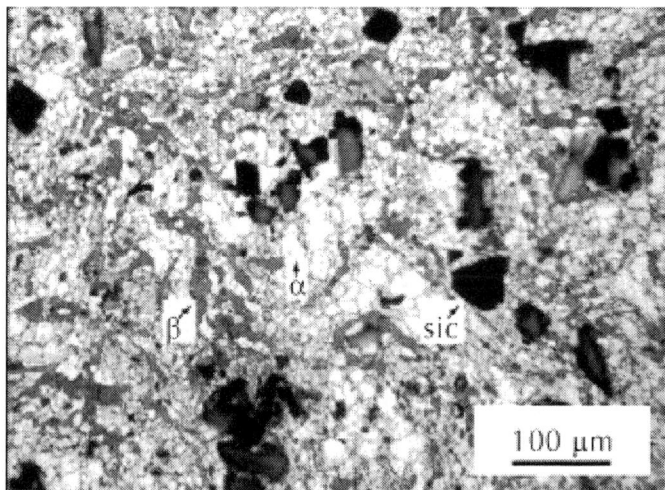

Figure 2: SEM micrograph of Mg8Li/SiC.

Preparation of Mg Based Nanocomposites and Ultrafine Grained Materials

Microcrystalline magnesium (μ-Mg) with 3 vol.% of alumina (Al_2O_3) nanoparticles (np) (μ-Mg+3nAl_2O_3), μ-Mg with 3 vol.% of zirconia np (μ-Mg+3ZrO_2) and microcrystalline Mg (μ-Mg) were studied. Micro-scaled Mg powder having a particle diameter of about 40 μm was prepared by gas atomisation of a high purity Mg melt in Ar atmosphere containing 1% oxygen for powder passivation. Both alumina and zirconia powders with a mean particle size of 14 nm were prepared by evaporation with the pulsed radiation of a 1000 W Nd:YAG laser and subsequent condensation of the laser-induced vapor in a controlled aggregation gas. The preparation method of np is described elsewhere (Naser et al., 1997, Ferkel & Mordike, 2001). The Mg powder was mixed with ceramic np in an asymmetrically moved mixer for 8 h. The powder mixture was then milled together for 1 h in a planetary ball mill. Mixture was subsequently pre-compressed followed by hot extrusion at a temperature of 350 °C under a pressure of 150 MPa. After extrusion, the originally more or less equiaxial grains changed into elliptical grains with the long axis parallel to the extrusion direction. The grain size in the cross section was about 3 μm and in the extrusion direction 10 μm. The distribution of the np was not homogenous. The np or their agglomerates were located mainly along the grain boundaries of the resultant material. The transmission electron microscopical inspection revealed that only few particles were distributed within grains. Similar procedure was used for preparation of μ-Mg with 3 vol.% of alumina microparticles (μ-Mg+3μAl_2O_3)

Details of the preparation of the UFG-Mg and nanocomposite (nc) with the 3 vol.% of graphite nanopartiles (Gr np) (UFG-Mg+3nGr) was similar to the nc with the ceramic np. The microscaled Mg powder was prepared by gas atomisation of a high purity magnesium melt with argon atmosphere containing 1% oxygen for powder passivation. The Mg-powder had a median particle diameter of about 40 μm. The graphite powder used had a median particle size of 1-2 μm. The Mg powder was mixed with 3 vol.% of graphite powder in an asymmetrically moving mixer for 8 h. The powder mixtures were then milled for 8 h at 200 rpm in the planetary ball mill (Retsch, PM400) in a sealed argon atmosphere. The milling vessel of 500 ml volume was made of corundum and the milling balls (11 mm diameter) were made of hardened steel (100Cr6). The weight ratio of ball-to-powder was 10:1. The composite was encapsulated in an evacuated Mg container (70 mm in diameter), degassed at 350 °C, and extruded by the preheated (350 °C) 400 t horizontal extrusion press (outlet 14 mm). Analyses of the extruded material in an optical spark analyser (Spectrolab, Spectro Analytical Instruments) reveal no contamination of the composite by e.g. Fe from the extruder tools or milling balls, or Al from milling vessel (Ferkel, 2003). The mean grain size of specimens used was estimated, using transmission electron microscopy and X-ray profile analysis, to be about 150-200 nm. TEM of the UFG-Mg$^+$3nGr is shown in Fig. 3.

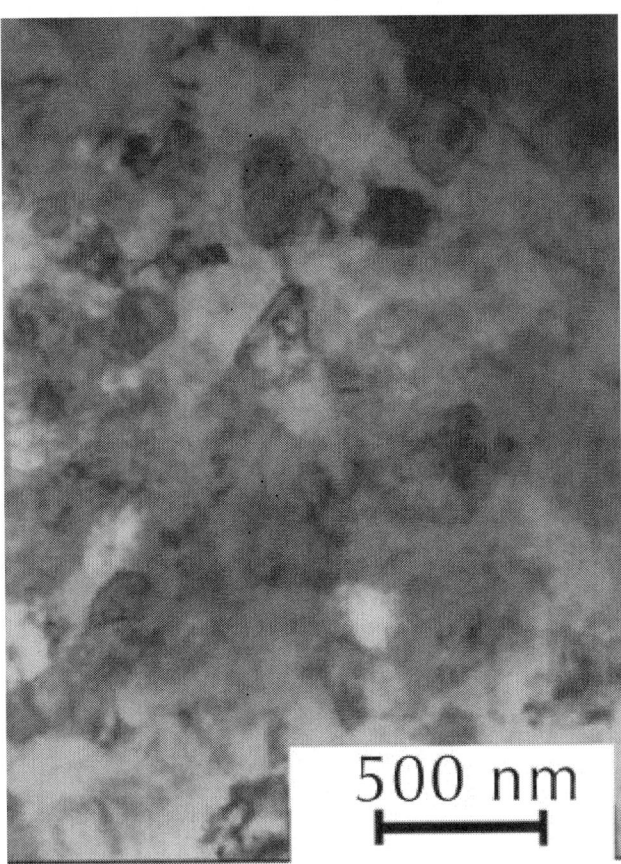

Figure 3: TEM of UFG-Mg with 3vol% Gr.

MECHANICAL PROPERTIES

Composites Reinforced With SiC Microparticles

Compressive Stress-Strain Curves

Deformation tests were carried out at temperatures between room temperature and 450 °C using an INSTRON testing machine. Cylindrical specimens of 8 mm diameter and 12 mm length were used for deformation tests performed in compression. The tensile specimens exhibited a gauge length of 10 mm and gauge diameter of 6 mm. Both samples were machined with a stress axis parallel to the extrusion direction. The strain rate sensitivity parameter *m* has been estimated by the abrupt strain rate changes (SRC) method. SRC tests and tensile tests with a constant strain rate ($\dot{}$ = *const.*) were performed at temperatures from 300 to 450 °C. Temperature in the furnace was kept with an accuracy of ±1°C. Fig. 4 shows the compressive true stress-strain curves obtained for AZ91/SiC composite deformed at various temperatures. Samples were deformed either to fracture or at higher temperatures to predetermined strains. The stress-strain curves obtained at temperatures up to 150 °C exhibit the small hump at the beginning of deformation. Curves at temperatures higher than 150 °C are very flat; the maximum stress is achieved at lower strains. Similar stress strain curves obtained in compression for WE54+13%SiC composite at various temperatures are introduced in Fig. 5. The stress-strain curves obtained at higher temperatures have a flat character. The temperature influence on the strain hardening of the composite is well visible. The yield stress decreases with increasing temperature very slowly up to 200 °C. At temperatures higher than 200 °C both characteristic stresses (the compression yield stress and the compression strength) decrease substantially. It can be concluded that the thermal stability of the composite is up to 200 °C very good. The stress-strain curves estimated for Mg-8Li/SiC composite at various temperatures are shown in Fig. 6. A continuous decrease in characteristic stresses can be seen from Fig. 6.

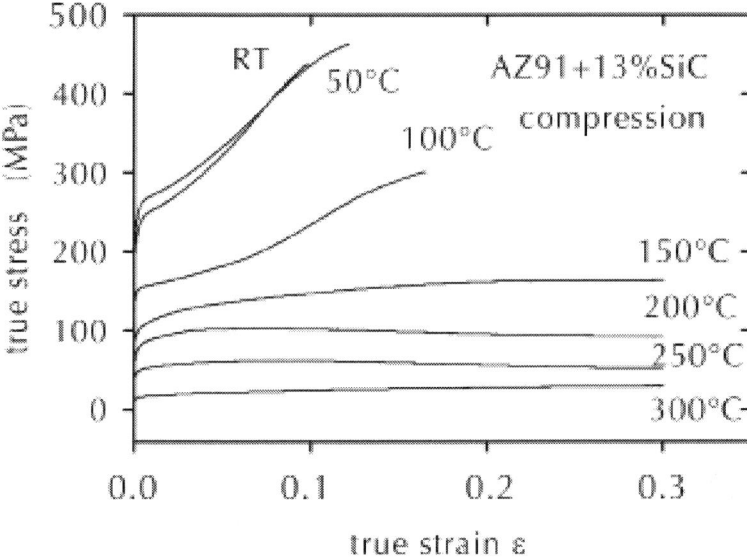

Figure 4: Compressive true stress-true-strain curves obtained for AZ91/SiC samples.

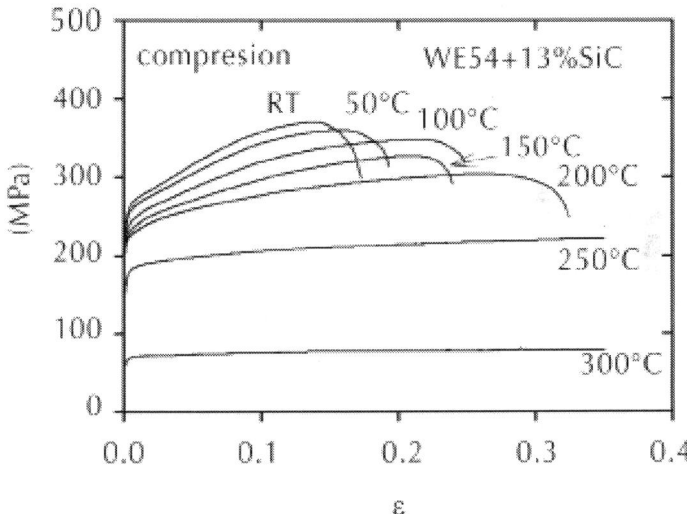

Figure 5: Compressive true stress-true strain curves obtained for WE54/SiC samples.

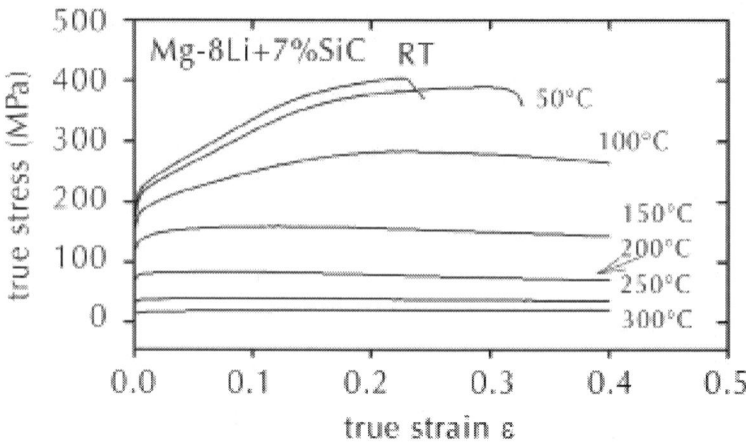

Figure 6: Compressive true stress-true strain curves obtained for Mg8Li.

The compression yield stresses (CYS) estimated as the proof stresses at a strain of 0.002 are introduced in Table 1 together with the ultimate compression strength values (UCS). Three characteristic temperatures have been chosen.

Table 1: Values of the yield stress and the ultimate compression strength estimated for three temperatures

	RT		150 °C		300 °C	
	CYS (MPa)	UCS (MPa)	CYS (MPa)	UCS(MPa)	CYS (MPa)	UCS (MPa)
AZ91/13SiC	222.5	437	98	163.8	15.6	30.6
WE54/13SiC	245.6	370	219.7	327.4	64.5	81.6
Mg8Li/7SiC	198.3	404.6	126.5	159	14.8	22

Table 1 shows a rapid decrease in the characteristic stresses obtained for AZ91/SiC and Mg8Li/SiC composites, while the decrease in the case of WE54/SiC composite is not so significant. This behaviour is very probably due to the presence of rare earth elements in the alloy. Rare earth elements form thermally stable precipitates situated in the grain boundaries. These grain boundary networks are also the reason for the good creep resistance.

TEM of the non-deformed WE54 composite sample and the WE54/ SiC sample deformed in compression at 50 °C are shown in Figs. 7 and 8, respectively. TEM investigation showed that presence cuboidal particles which were identified as $Mg_{12}NdY$ precipitates. Many twins are a common feature for both non-deformed as well as deformed material. Grains in as-received material are well visible in Fig. 8. Thin twins, within single grains, are parallel to each other and quite narrow. Sometimes twins are extended through the grain boundary, causing a grain boundary deflection. The role of twinning is well known in deformation of *hcp* lattice alloys. Twinning is an important deformation mode of Mg alloys. It reorients the slip planes in order to relax stress concentrations and enhances multiple slip. Thin twins, often appearing in the form of parallel groups, where detected in compressed material (Ion et al., 1982). Another feature in the microstructure of the deformed composites is a high dislocation density. Dislocations in many cases form pile ups and tangles (see Fig. 8). Significant amount of rectangular shape fine particles was estimated distributed within the grains.

Figure 7: Twins and precipitates in non-deformed composite. Dislocations in the right bottom corner are visible.

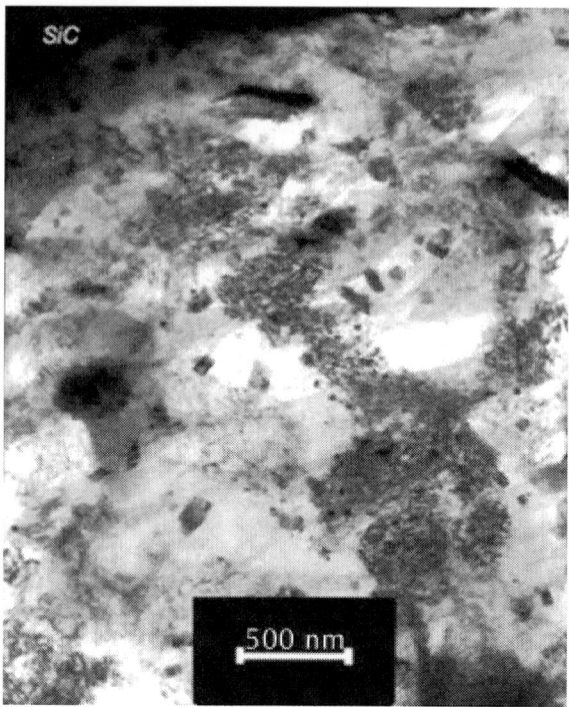

Figure 8: TEM of the WE54/SiC sample deformed at 50°C. SiC particle is situated in the top left corner of the picture.

In order to obtain more detailed information concerning the deformation mechanism(-s) occurring in particle reinforced magnesium alloys based composites, acoustic emission (AE) was used. AE stems from transient elastic waves which are generated within the material during deformation due to sudden localized and irreversible structure changes like dislocation glide and twinning, which may be considered as the main deformation mechanisms in Mg and its alloys due to their hexagonal crystal structure. WE54/SiC samples were deformed in compression at room temperature. The engineering stress- time plots together with the time variation of the AE count rates are shown in Fig. 9. Two AE maxima were observed. The observed AE maximum at the onset of plastic deformation is connected with the yield point (CYS $\cong$ 280 MPa) and it may be ascribed to the stochastic $\{10\bar{1}2\}$ $\langle 10\bar{1}0 \rangle$ primary twin formation in the grains unfavourably oriented for the basal slip during the very early stage of plastic deformation. These twins reorient the original lattice on 86.3° and the subsequent straining may continue by the basal slip and secondary twinning.

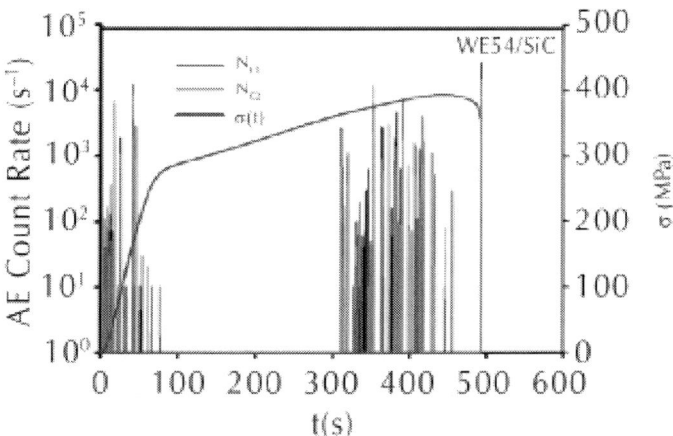

Figure 9: Deformation curve and AE count rate for WE54/SiC composite obtained at ambient temperature.

New twins were observed in the deformed microstructure (Fig. 10). Slim twins were observed also by the TEM as demonstrated in Fig. 8. These twins (and also dislocations visible in Fig. 8) arose very probably during extrusion of the powder material. Possible dislocation contribution to the AE signal is very probably marginal due to small grain sizes. The slip length of dislocations is very short (see dislocation tangles in the left bottom corner of the micrograph 8) and therefore the probability of pile ups formation is also very low. The mobility of dislocations is further limited due to the presence of small cuboidal particles visible in Fig. 7 and 8. The AE signal detected at the time point of ~310 s and onwards is discontinuous, and its sources may be ascribed to damage processes. Specifically, with the addition of the reinforcing phase, the geometrically necessary dislocations are generated to accommodate the plastic mismatch in the matrix. The stress concentrations in the vicinity of the reinforcing SiC particles may achieve their critical value and a breakage of particles and/or a release of cracks (decohesion) between the matrix and particles can occur. Both processes are considered as strong sources of the AE signal. The broken SiC particles and also cracks in the vicinity of particles were observed at the polished surface of the sample deformed to failure (see Fig. 11). Two main mechanisms were detected during plastic deformation: breakage of particles and decohesion in the particle—matrix interface. Based on these results we may consider localisation of the plastic deformation and fracture of the sample if the number of broken particles achieves its critical value.

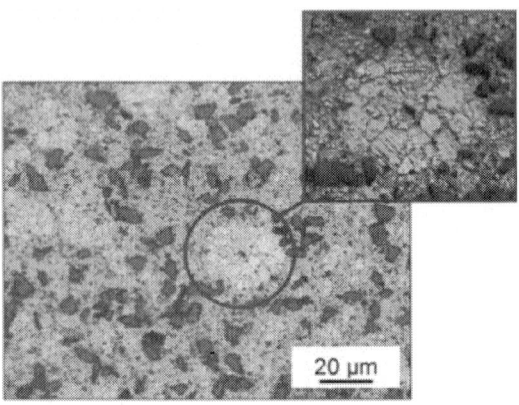

Figure 10: Light micrographs showing new twins formed during plastic deformation in the WE54/SiC sample.

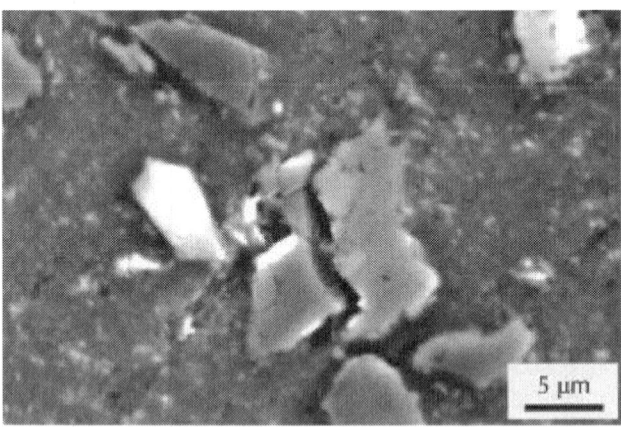

Figure 11: Scanning electron micrograph showing broken particles on the surface of the deformed WE54/SiC sample. Decohesion between the matrix and particles is also visible.

High Temperature Properties Estimated in Tension

The tensile true stress−strain curves estimated for WE54/SiC samples at a strain rate of 2.8×10^{-4} s^{-1}and various temperatures are shown in Fig. 12. A significant work hardening was obtained at temperatures up to 150 °C. A considerable decrease in the flow stress was observed after increasing the temperature from 250 to 300 °C. Curves obtained at temperatures higher than 250 °C have a flat character. It indicates new deformation process(-es) taking place at temperatures higher than 250 °C. Small grain sizes of powder metallurgically prepared materials indicate possibility of superplastic deformation. To check this eventuality, SRC tests were performed at temperatures from 350 to 450 °C. The values of the strain rate sensitivity m estimated for various strain rates and temperatures are given in Fig. 13. The strong strain rate dependence of the m-parameter is obvious from the picture. With increasing temperature the dependence is shifted to higher strain-rates. However, the m parameter slightly increases with temperature; the maximum values are in the vicinity of 0.3 as it is obvious from Table 2.

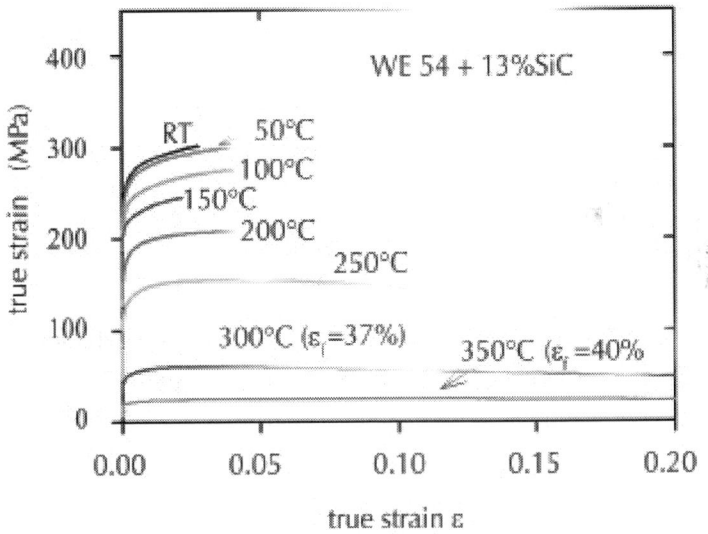

Figure 12: Tensile true stress-true strain curves estimated for WE54/SiC composite at various temperatures.

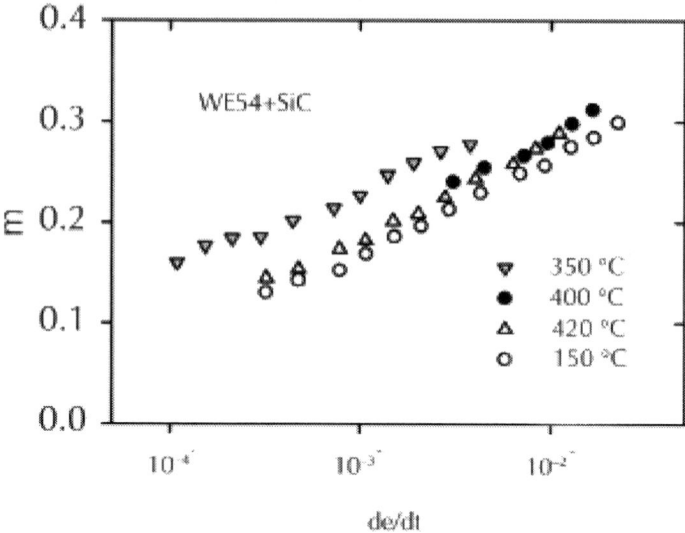

Figure 13: Strain rate dependence of m-parameter estimated at various temperatures.

The maximum recorded elongation to failure was 99%, which shows an evidence of the enhanced plasticity, nevertheless this value remains below the bottom limit for superplastic region. SEM micrograph of the sample exhibiting the highest elongation is shown on Fig. 14 documenting the microstructure after deformation at 450 °C. Numerous cavities formed during the high temperature deformation are visible in Fig. 14. The cavitation primarily occurred at the reinforcement/matrix interfaces, which are the preferential sites for the nucleation of cavities. The density of cavities was higher near the fracture surface where elongated cavities were found. Since many of these cavities were fairly large, it is reasonable to assume that growth and subsequent coalescence and interlinkage of the cavities led to the premature failure.

Figure 14: SEM micrograph showing cavities formed during high temperature deformation.

Table 2: Maximum values of m-parameter and elongation to failure estimated for three temperatures and corresponding strain rates

	x_{mma}	$\dot{}(s^{-1})$	$_f(\%)$
350 °C	0.28	4×10^{-3}	66
420 °C	0.30	2×10^{-2}	95
450 °C	0.29	5×10^{-2}	99

Similar study was performed for the Mg8Li/SiC samples. The strain rate sensitivity values m, obtained in tension using SRC method at temperatures 200, 250 and 300 °C are introduced in Fig. 15. It is obvious that the m−values increased with increasing deformation temperature and the dependence was shifted to the higher strain rates. While at 200 °C the maximum of m was reached at $\dot{} = 4 \times 10^{-5}$ s^{-1}, at a temperature of 300 °C the maximum laid in the vicinity of the strain rate $\dot{} = 6 \times 10^{-4}$ s^{-1}. The maximum value of m estimated at 300 °C exhibited 0.46, which is close to a value of 0.5, considered as an optimal value for the superplasticity. The maximum value of $m = 0.3$ at 200 °C and all values of m estimated at temperatures higher for all strain rates were in the region enclosing the superplastic behaviour. The maximum ductility $A = 110\%$ was found at 300 °C and at a strain rate of 6×10^{-4} s^{-1}, which correspond to the maximum of the strain rate sensitivity $m = 0.46$. Despite of relatively high values of the strain rate sensitivity m, the achieved ductility was only at the onset of the superplastic region.

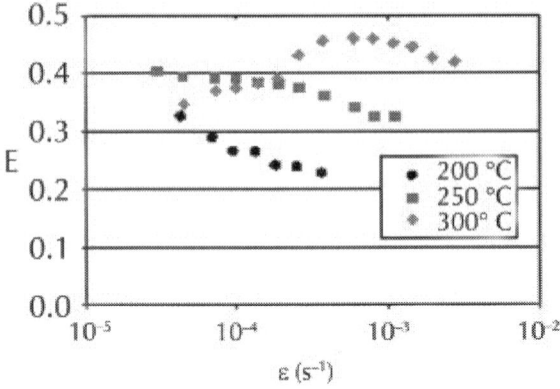

Figure 15: Strain rate dependence of the strain rate sensitivity parameter estimated for Mg8Li/SiC.

The observed cavities formation after high temperature deformation in WE54/SiC and Mg8Li/SiC composites indicates the presence of some diffusion process. The activation energy Q of such rate controlling is given by the relationship (Langdon, 1994):

$$\dot\varepsilon = \frac{AD_0Gb}{kT}\left(\frac{b}{d}\right)^{p}\left(\frac{\sigma}{G}\right)^{n}\exp\left(-\frac{Q}{RT}\right), \tag{1}$$

where $\dot{}$ is steady−state deformation rate, A is a dimensionless constant, d is the grain size, σ is the applied stress, p is the grain size exponent, n = 1/m is the stress exponent and D_0 is the pre−exponential factor; kT has its usual meaning and R is the gas constant. The activation energy Q is a slope of the plot σ/G vs $1/T$ (when $\dot\varepsilon T/G$ is constant):

$$Q = \frac{1}{m}R\frac{\Delta\left(\ln\sigma/G\right)}{\Delta\left(1/T\right)} \tag{2}$$

The activation energy was estimated using relationship (2) for both composites to be Q(WE54/SiC) = 114kJ/mol. The activation energy for the volume diffusion is 135 kJ/mol and for the grain boundary diffusion 92 kJ/mol (Frost & Ashby, 1982). Comparing to an experimental value of 114 kJ/mol (using rule of mixture), we may conclude that the measured activation energy consists of approximately 50% volume and 50% grain boundary diffusion. Using estimated values of m, the activation energy for Mg8Li/SiC composite was found (for the high m region) Q ≈ 87 kJ/mol. According to binary Mg-Li diagram (Nayeb-Hashemi et al., 1984), the Mg−8Li alloy at 300 °C consists of equilibrium   phase and   phase whose chemical composition are 5.7 Li and 11 Li, respectively. Owing to higher mobility of Li in   phase the Li content in   phase decreases while in the   phase increases and the volume fraction of the   phase increases at the expense of   phase. Due to grain boundary migration caused by atomic mobility, the grain size of   phase increases.

Experimentally estimated value of the activation energy Q = 87 kJ/mol indicates that the main rate controlling mechanism is the grain boundary sliding accommodated with the grain boundary diffusion with the small contribution of the lattice diffusion. The successive grain growth increases the grain size in the   phase. The diffusion accommodation of the grain boundary sliding is more difficult, which implies cavities formation.

Observed formation and growth of cavities relaxes the stress concentration caused at the particles on the sliding grain boundaries. Cavities, created by vacancy clustering, may nucleate if the stress concentration is not relieved sufficiently rapidly. Local tensile stress caused by sliding at interfaces may be written in the form (Mabuchi & Higashi, 1999):

$$\sigma_{slid} = \frac{0.92kTd_p\dot{\varepsilon}dV_f}{\Omega D_L\left(1+5\dfrac{\delta D_{GB}}{d_p D_L}\right)} \qquad (3)$$

where dp is the particle diameter, ˙ is the strain rate, d is the grain size, DL is the lattice diffusion and DGB is the grain boundary diffusion coefficient,   is the grain boundary width, Ω is the atomic volume. Vf is the volume fraction of particles and kT has its usual meaning. The insufficiently accommodated grain boundary sliding process is the reason for cavitation and early failure of samples.

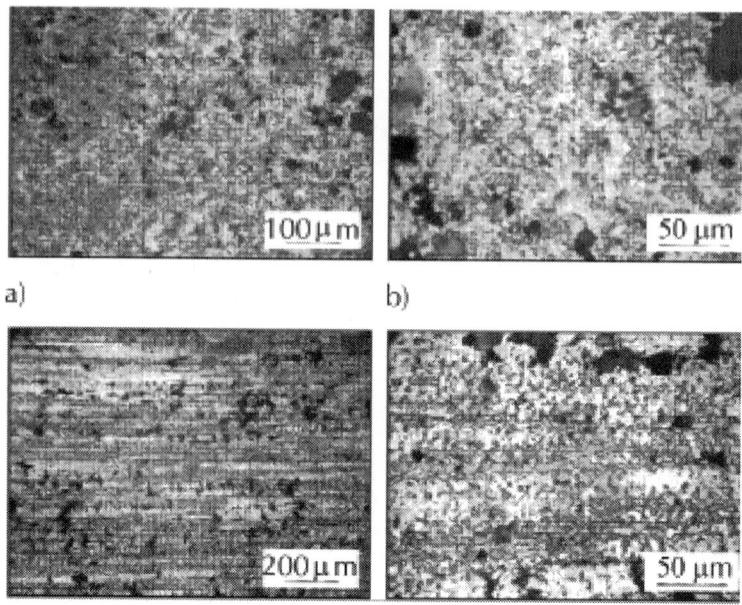

a)   b)

Figure 16: Light micrograph of Mg8Li/SiC composite after deformation at 300 °C and strain rate 6×10^{-4} s^{-1}, a), b) cross section, c), d) longitudinal section.

Based on our results of mechanical tests, microstructural observations and the analysis of the AE signal occurring during plastic deformation, we may conclude that the deformation processes in the powder metallurgically prepared magnesium alloys based composites are different at lower and higher temperatures. At lower temperatures (below about 150 °C), the deformation processes have following main characteristics:

i.   Small grain size – several micrometers - is a typical feature of the microstructure.

ii.   Early stages of the compressive plastic deformation are realised by twinning accommodated with the dislocation glide;

iii.   Twin boundaries are impenetrable obstacles for the dislocation motion and contribute to a significant hardening at lower temperatures;

iv. Breakage of particles and decohesion in the particle–matrix interfaces cause localisation of plastic deformation and failure of materials;

Although the grain size of composites exhibiting 3-5 µm indicated a possibility of the occurrence of the structural superplasticity, only enhanced plasticity was estimated at higher temperatures. This fact has following reasons:

1. The values of the strain rate sensitivity parameter reached $m = 0.3\text{-}0.5$;

2. The grain boundary sliding is the significant mechanism operating at higher temperatures;

3. The stress concentrations formed at the particles on the sliding grain boundaries are the reason for the cavity formation.

4. Growth and subsequent coalescence and interlinkage of the cavities led to the premature failure of composites.

Stress-Strain Curves of Magnesium Micro- and Nanocomposites

Microcrystalline Magnesium Reinforced With Ceramic Microparticles and Nanoparticles

True stress-true strain curves obtained for microcrystalline Mg reinforced with alumina microparticles are introduced in Fig. 17a for tension and 17b for compression. A substantial difference between shapes of curves measured in tension and compression are visible. Similar curves were obtained also for microcrystalline Mg reinforced with alumina and zirconia np as it follows from Figs 18a,b and 19 a,b. It can be seen that the flow stress decreases as the test temperature is increasing. Substantial differences between curves obtained in tension and compression are obvious comparing Fig. 18a,b and19a,b. While the curves measured in tension are mostly flat (for nc with alumina np at a temperature of 100 °C and higher), the curve estimated in compression exhibit local maxima: for $\mu\text{-}Mg+3Al_2O_3$ at 100 °C and for $\mu\text{-}Mg+3ZrO_2$ at room temperature and 100 °C. The main deformation mode in magnesium is the basal slip – glide of <a> dislocations. The secondary conservative slip may be realised by the motion of <a>–dislocations on prismatic and pyramidal planes of the first order. The basal <a> dislocations may react with the pyramidal <c+a> dislocations. Different dislocation reactions may produce both sessile and glissile dislocations (Trojanová et al., 2011). Production of sessile dislocations increases the density of obstacles for moving dislocations. Cross slip of <a> dislocations on prismatic planes and climb of <c> dislocations are the main dynamic recovery mechanisms. The flat course of the stress-strain curves is a result of a dynamic balance between hardening and softening. To fulfill the von Mises criterion for compatible deformation of polycrystals, the non-basal <c+a> slip and /or deformation twinning are needed (Yoo et al., 2001). In coarse grained Mg and Mg alloys, twinning is observed in the early stages of plastic deformation.

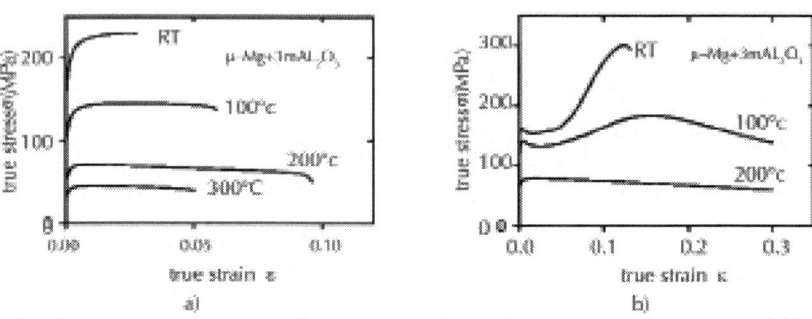

Figure 17: True stress-true strain curves obtained for μ-Mg+3%mAl$_2$O$_3$ in tension (a) and compression (b).

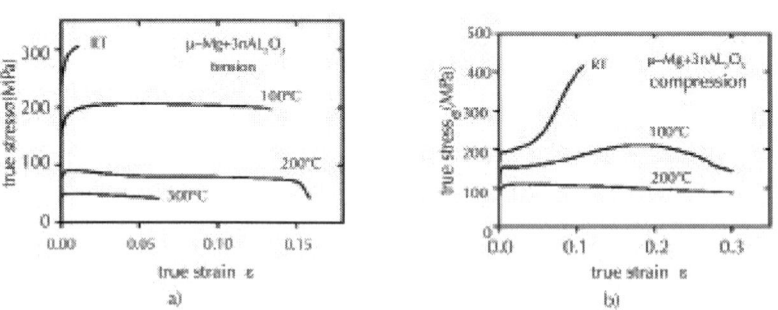

Figure 18: True stress-true strain curves obtained for μ-Mg+3%nAl$_2$O$_3$ in tension (a) and compression (b).

The stress-strain curves obtained in tension are flat at temperatures above 100 °C, there is a dynamic balance between hardening and softening. The work hardening rate is very close to zero. Tensile/ compression yield stress asymmetry was studied by Mathis et al. using acoustic emission technique (Máthis et al., 2011). A broad maximum on the curves estimated at room temperature and at 100 °C was observed at a strain of about 10 %. Analysis of the acoustic emission signal revealed that during the compression tests below 200 °C, the maximum of the twin formation is at a very beginning of the stress-strain curve. The twin boundaries are obstacles for dislocation motion. Therefore, twin growth caused an increase in the flow stress, which was manifested by a rapid decrease in the acoustic emission signal. In tension, a significant twin formation was observed during the entire test. Twinning mode of deformation requires formation of the new interface inside of grains. The energy of twin interfaces in Mg is relatively high (Koike et al., 2003). This leads to an increasing difficulty of twin nucleation with decreasing grain size. This has been confirmed by the acoustic emission measurements (Trojanová et al., 2008).

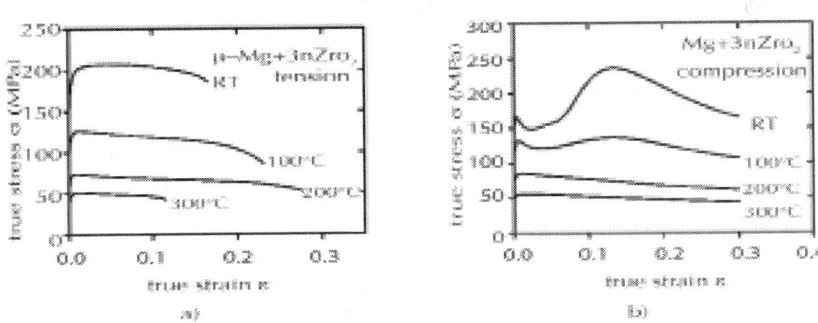

Figure 19: True stress-true strain curves estimated at various temperatures for.μ-Mg+3nZrO$_2$ sample in tension (a) and compression (b).

Table 3: The tensile yield stress and ultimate tensile strength estimated at various temperatures in tension

T	TYS (MPa)				UTS (MPa)			
°C	μ-Mg	μMg+ 3mAl$_2$O$_3$	μMg+ 3nAl$_2$O$_3$	μMg+ 3nZrO$_2$	μ-Mg	μMg+ 3mAl$_2$O$_3$	μMg+ 3nAl$_2$O$_3$	μMg+ 3nZrO$_2$
20	229.0	200.0	269.0	175.5	256.5	230.0	306.2	207.5
100	145.7	128.2	172.6	114.8	161.2	145.4	205.5	125.8
200	77.6	65.9	86.0	68.4	84.5	71.5	91.1	73.2
300	48.5	41.5	48.5	46.2		45.3	50.3	51.0

The temperature variations of the TYS and the UTS estimated for μ-Mg and μ-Mg with ceramic mp and np are introduced in Table 3. Relatively high values of the TYS and UTS were estimated. This is also influenced by small grain size of all materials as a result of the powder metallurgical preparation route. The Influence of the particles on the yield stress depends on the size of particles and kind of the np and, of course, on the test temperature. There are large differences in the values of the TYS at room temperature and 100 °C. The difference in the values of the TYS for different composites measured at room temperature may be explained by the presence of the np and by the bonding between the particles and the matrix. The difference in the values of the TYS measured at 200 °C is small and at 300 °C, the values of the yield stress are the same for μ-Mg and μ-Mg reinforced with ZrO$_2$ and Al$_2$O$_3$ particles. Relatively high difference between the TYS and the UTS of μ-Mg+3Al$_2$O$_3$ and μ-Mg+3ZrO$_2$ (approximately 100 MPa) may be caused by the strength of the bonding between Mg matrix and ceramic np. While this bonding between Mg and Al$_2$O$_3$ np is nearly perfect, in the case of ZrO$_2$ is weak as it will be demonstrated in the paragraph 4.1.

Table 4: The tensile yield stress and ultimate tensile strength estimated for various temperatures in compression

T	CYS (MPa)			UCS (MPa)		
°C	μMg +3mAl$_2$O$_3$−	μMg +3nAl$_2$O$_3$	μMg +3nZrO$_2$	μMg +3mAl$_2$O$_3$	μMg +3nAl$_2$O$_3$	μMg +3nZrO$_2$
20	162.8	188.9	162,1	300.8	415.7	236.4

100	135.2	140.8	124.7	183.7	210.6	136.2
200	72.2	96.0	79.6	79.0	110.5	85.0
300			52.2			56.9

The temperature variations of the CYS and the UCS are introduced in Table 4 for Mg reinforced with various types of particles. It can be seen that while the values of the CYS of all materials studied are lower, the UCS values are higher than those measured in tension. The twinning mode of deformation in the early stages of deformation is easier in the compressive straining. New interfaces formed inside the grains are impenetrable obstacles for dislocation motion and they lead to hardening which is manifested with the local maximum on the stress-strain curve. The height of this maximum estimated at RT and 100 °C agree with the UCS value. The role of twinning decreases with increasing deformation temperature.

Ultrafine Grained Magnesium

Figure 20a shows the true stress-true strain curves for UFG-Mg deformed in compression at various temperatures. Samples were deformed either to fracture or at higher temperatures to predetermined strains. A pronounced upper yield point followed by the hardening stage was observed at temperatures between room temperature and 100 °C. The stress-strain curves at temperatures higher than 150 °C have a steady state character. Stress-strain curves measured for UFG-Mg with 3 vol. % of Gr np (UFG-Mg+3nGr) in compression are presented in Fig. 20b for various temperatures. The local maximum at the beginning of the stress strain curve was in this case observed only at ambient temperature. The temperature dependences of the CYS and UCS estimated for both materials are shown in Table 5. Both stresses decrease rapidly with increasing temperature. The yielding phenomenon is very probably caused by the avalanche release of dislocations pinned at grain boundaries. It will be shown later in the paragraph 4.2 that the amplitude dependence of the logarithmic decrement in UFG-Mg indicates this phenomenon. Acoustic emission measurements detected no signal from the deformed sample at room temperature. This indicates that no dislocation pile-ups have been formed. In order to produce a pile-up of dislocations the grains need to contain a Frank-Read source operating in a cyclic way.

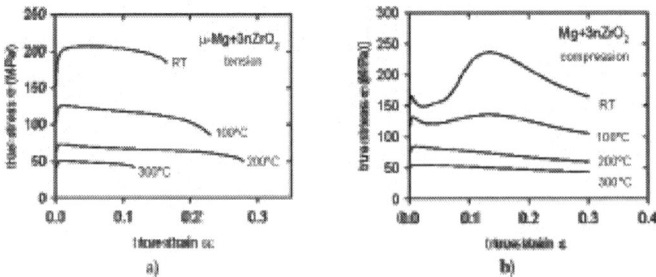

Figure 20: True stress-true strain curves obtained for various temperatures in compression for UFG-Mg (a) and UFG-Mg+3nGr (b).

Table 5: The compressive yield stress and ultimate compression strength estimated for various temperatures for UFG-Mg and UFG_Mg with Gr np

Temperature	UFG-Mg compression		UFG-Mg+3Gr compression	
T (°C)	CYS (MPa)	UCS (MPa)	CYS (MPa)	UCS (MPa)
22	222.0	335.9	259.4	293.2
50	209.3	300.3		
77			196.3	212.0
100	170.2	208.3		
150	124.0	135.1		
200	89.5	95.7	149.3	157.9
250	59.6	76.1		
300	59.6	68.0	111.0	129.4

The maximum link length of a source to operate in a cyclic way becomes about D/2 (D is the separation of grain boundaries). The minimum resolved shear stress on the grain boundary due to the first loop can be written as

$$\tau_{GB} = \frac{2Gb}{D} - \frac{Gb}{D} = \frac{Gb}{D} \tag{4}$$

where G is the shear modulus and b the Burgers vector of dislocations. The first term is the operating stress of the source and the second is the back stress from the first loop that is forced up against the grain boundary. If $_{GB}$ exceeds the critical stress $_{c'}$ required to an activation of an accommodating system, the first loop will escape in the grain boundary before the second loop is ejected from the source, and no pile-up will be formed (Nes et al., 2005). The stress-strain curves at temperatures higher than 150 °C exhibit a steady state character indicating a dynamic balance between hardening and softening.

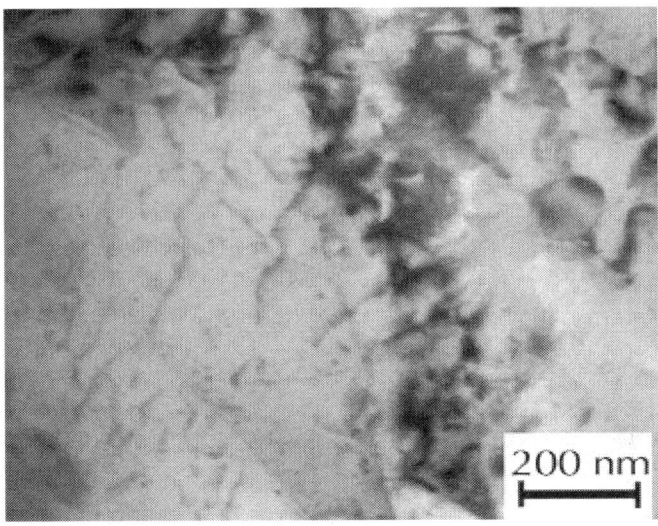

Figure 21: TEM of UFG-Mg+3nGr deformed at room temperature.

It is clear that the Gr np increase the CYS. It is interesting to note that the UCS value of UFG-Mg deformed at RT is higher than that of Mg+3nGr. Values estimated at 300 °C for both materials show that the softening in the Mg+3nGr is lower than that in UFG-Mg. In the case where materials were prepared by milling, the refinement of the originally large grains into the UFG region is connected with large plastic deformation. This large plastic deformation is partially recovered during consolidation of the powder. Such mc and UFG materials usually contain a high density of dislocations in a heavy-deformation micro-structure (Mohamed, 2003). Np strengthening and grain refinement of the matrix are the key strengthening mechanisms in the UFG-Mg+3nGr nc (Luká et al., 2006, Estrin & Vinogradov, 2013) A TEM micrograph of the UFG-Mg+3nGr sample deformed at RT is shown inFig. 21. Dislocations pinned at grain boundaries are visible.

The influence of grain size on the flow stress of magnesium alloys has been subject of a number of investigations (e.g. Mabuchi et al., 2001, Anderson et al., 2003). Choi and co-workers (Choi et al., 2010) studied the deformation behaviour of ball milled Mg with the grain size from 60 nm up to 1 μm. They estimated that the YS is result of many mechanisms including the Hall-Petch strengthening, deformation twins, emission of partial dislocations and grain boundary sliding. Based on these estimations the deformation modes in the extruded Mg can be at RT divided into three regions:

1. Mc region (the grain size above 1 μm): the yield stress of Mg follows the Hall-Petch relationship.
2. UFG region (the grain size between 100 nm and 1 μm): the YS of Mg negatively deviates from the Hall-Petch relationship.
3. Nc region (the grain size smaller than 100 nm): the slope of the yield stress dependence on the grain size is negative.

While for mc Mg the interaction of dislocations with grain boundaries and/or twin boundaries can be considered as significantly affecting plastic deformation, for UFG-Mg the dislocation pile-up mechanism loses its significance; twining and grain boundary sliding gradually contributes to plastic deformation.

X-ray Analysis

Deformed UFG-Mg samples were investigated by the high resolution X-ray diffraction peak profile analysis. The diffraction profiles are evaluated by assuming that peak broadening is caused by small crystallites and strain caused by dislocations. The Williamson–Hall plots (integral breadth vs. sin , where is the Bragg angle) were constructed. The measured physical profiles are fitted by theoretical profiles calculated on the basis of well-established profile functions of size and strain. Both strain and strain anisotropy are accounted for by the dislocation model of lattice distortions (Wilkens, 1970; Ungár & Borbely, 1996). In a crystal containing dislocations, the mean square strain is done by Wilkens (1970)

$$\left\langle \varepsilon_{g,L}^2 \right\rangle \cong \left(\rho C b^2 / 4\pi \right) f(\eta) \tag{5}$$

where   is the density of dislocations and b is their Burgers vector, C is the dislocation contrast factor, f( ) is the Wilkens function, where = L/R$_e$ (R$_e$ is the effective outer cut-off radius of dislocations), L is the Fourier length defined as L = na$_3$ (a$_3$ = /2(sin $_2$-sin $_1$)), n are integers starting from zero, is the wavelength of X-rays and ($_2$- $_1$) is the angular range of the measured diffraction profile. In the case of hexagonal polycrystals, where dislocations with various Burgers vectors take place in deformation, the average contrast factor may be obtained by the weighted linear combination of the individual contrast factors for the active sub slip systems:

$$\left\langle Cb^2 \right\rangle = \sum_{i=1}^{N} f_i \left\langle C^i \right\rangle b_i^2 \tag{6}$$

where N is the number of the different activated sub slip systems, $\langle Ci \rangle$ is the average dislocation contrast factor corresponding to the ith sub slip system and f$_i$ are the fractions of the particular sub slip systems by which they contribute to the broadening of a specific reflection. The average contrast factors for a single sub slip system (hk.ℓ) in hexagonal crystals are (Dragomir & Ungár, 2002):

$$\left\langle C_{hkl} \right\rangle = \left\langle C_{hk.0} \right\rangle \left[1 + q_1 x + q_2 x^2 \right] \tag{7}$$

where x = (2/3)(ℓ/ga)^2, q$_1$ and q$_2$ are parameters depending on the elastic properties of the material, C$_{hk.0}$ is the average contrast factor corresponding to the hk.0 type reflections, g is the absolute value of the diffraction vector and a is the lattice constant in the basal plane. The q$_1$ and q$_2$ parameters and the values of C$_{hk.0}$ have been calculated for hexagonal crystals and compounds in (Dragomir & Ungár, 2002). Two Burgers vectors types were taken into account: ⟨a⟩=13⟨2$^-$110⟩ and ⟨c+a⟩=13⟨2$^-$113⟩. Using the scheme described by Kužel & Klimanek (1989), it is possible to take for the number N of sub-slip systems in eq. (12) with the <a> or <c+a> Burgers vector N⟨a⟩ = 4 and N ⟨c+a⟩ = 5, respectively. Once the Burgers vector types are determined, the value of ⟨Chk.0b2⟩and the dislocations density   can also be calculated; for further details see (Kužel & Klimanek, 1989). The experimental values of q$_1$ and q$_2$ denoted as can be estimated by the whole profile fitting procedure (details see (Kužel, 1998)).

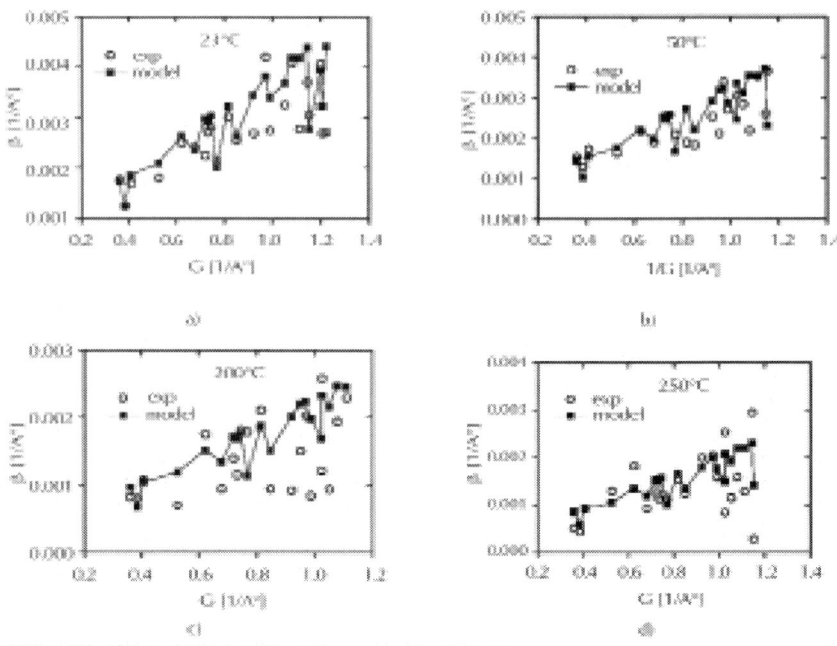

Figure 22: The integral breadths for the sample deformed at RT (a), 50 °C (b), 200 °C (c), 250 °C.

The Williamson–Hall plots in terms of the integral breadth vs. the magnitude of the reciprocal lattice vector G (G represents sin) are shown in Figs. 22 for samples deformed at various temperatures. Circles correspond to the values calculated from the experimental data (Kužel, 1998). Squares correspond to the values calculated on the basis of the model (Dragomir & Ungár, 2002). The fitting procedure allows to estimate the density of basal <a> dislocations, the density of <c+a> dislocations and the grain size depending on the testing temperature. The results are given in Table 6. From Table 6 it follows that the basal dislocation density decreases with increasing deformation temperature, while the fraction of <c+a> dislocations is constant. The grain size growth has been also observed with increasing deformation temperature. Máthis et al., (2004) used the similar method to study the evolution of the dislocation density with the Burgers vector of different types as a function of the testing temperature in a coarse grained Mg. They found that at room temperature and at 100 °C, the dominant dislocations are <a> or mainly basal dislocations.

Table 6: Basal dislocation density $_{<a>}$, density of <c+a> dislocations $_{<c+a>}$ and grain size estimated for various deformation temperatures

Deformation temperature (°C)	23	50	200	250
Density of basal dislocations $_{<a>}$ 10^{14} (m^{-2})	4.04	3.0	1.5	1.2
$_{<c+a>}/_{<a>}$	0.1	0.1	0.1	0.1
Grain size (nm)	150	200	350	400

At higher temperatures, the fraction of <a> dislocations decreases, whereas the fraction of <c+a> dislocations increases. The fraction of <c> dislocations remains practically unchanged. The average total dislocation density increased considerably with strain as compared to the value in the as-cast state.

However, with increasing deformation temperature, the increment of the dislocation density decreased strongly in the plastically deformed sample. The concomitant increase of the fraction of <c+a> dislocations and the decrease of the average dislocation density with increasing deformation temperature can be explained by the dynamic recovery mechanism/-s. The results obtained in our study for UFG-Mg exhibit similar decrease in the basal dislocation density. The fraction of the <c+a> dislocations in UFG Mg remains constant, while it increases in the coarse grained material. This is very probably caused by the different deformation mechanisms in the UFG-material.

INTERNAL FRICTION MEASUREMENTS

Amplitude Dependence of Internal Friction

The damping measurements were carried out in a vacuum (about 30 Pa) at room temperature. The specimens fixed at one end were excited into resonance by a permanent magnet fixed at the free side of the bending beam and a sinusoidal alternating magnetic field. The internal friction was characterised by the logarithmic decrement of the free decay of the vibrating beam $= \ln(A_n/A_{n+1})$, where A_n and A_{n+1} are the amplitudes of a free decay of vibrations after n and $(n+1)$ cycles, respectively. The resonance frequency ranged from 130 to 160 Hz. The strain amplitude dependences of the logarithmic decrement were measured. The specimens were annealed step by step at increasing temperatures up to 500 °C for 0.5 h and after each annealing cycle quenched into water of room temperature. The annealing at higher temperatures was performed in an argon atmosphere to avoid oxidation. The damping measurements were carried out immediately after heat treatment and quenching at room temperature.

Figure 23 shows the logarithmic decrement plotted against the logarithm of the maximum strain amplitude for μ-Mg+3nZrO$_2$ samples. The logarithmic decrement was measured before (as received - as rec) and after step by step annealing at increasing upper temperature of the cycle. It can be seen that the measured strain dependence of the logarithmic decrement exhibits two regions and can be expressed as

$$\delta(\varepsilon) = \delta_0 + \delta_H(\varepsilon) \tag{8}$$

where $_0$ is the amplitude independent component (or only weakly dependent on the maximum strain amplitude) found in the first region, for lower strain amplitudes. In the second region, for higher strain amplitudes, the component $_H()$ depends on the strain amplitude (amplitude dependent internal friction – ADIF); it increases with increasing strain amplitude. The upper temperature of the thermal cycling influences the logarithmic decrement. In the case of the μ-Mg+3nZrO$_2$ composite, the thermal treatment influences mainly the amplitude independent component of the decrement. It is important to note that very high values of damping in the amplitude independent component - in order of 10^{-2} - were obtained. The strain amplitude dependences of the logarithmic decrement for μ-Mg+3nAl$_2$O$_3$ are given in Figures 24. Figure 24a shows the strain amplitude dependence of the logarithmic decrement for specimens annealed up to 300 °C. Figure 24b shows the strain amplitude dependence of the logarithmic decrement for specimens annealed at temperatures between 300 and 500 °C. The thermal cycling influences mainly the amplitude dependent component $_H$ that increases with increasing upper temperature of the cycle, if the upper temperature is between room temperature and 300 °C. For the upper temperature between 300 and 500 °C, the $_H$ component decreases with increasing upper temperature.

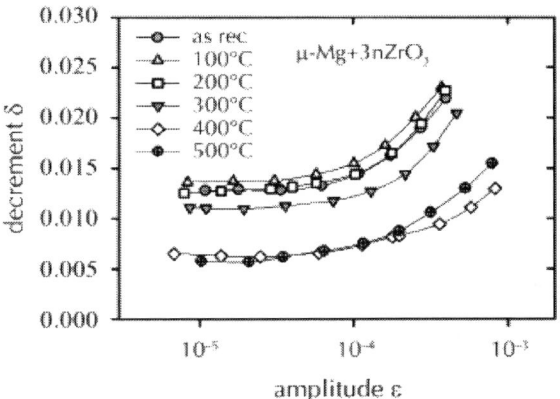

Figure 23: Amplitude dependence of decrement measured after annealing at increasing temperatures.

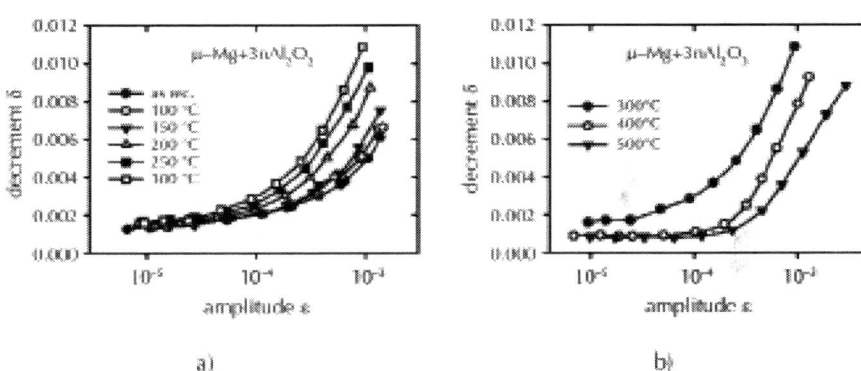

Figure 24: Amplitude dependence of the decrement for μ-Mg+3nAl$_2$O$_3$ annealed at temperatures up to 300°C (a) and temperatures between 300 and 500 °C (b).

The strain amplitude dependences of the logarithmic decrement measured in UFG-Mg after thermal cycling are given in Figure 25a.

Only a slight influence of thermal cycling on the   versus   curves can be observed. Figure 25b shows the plots of   versus   for UFG-Mg+3nGr. It can be seen that the thermal cycling causes a slight decrease of the logarithmic decrement at small strain amplitudes. The $_H$ component changes only very slightly with increasing temperature of the cycle. Figure 26 shows the strain amplitude dependence of the logarithmic decrement for various materials before annealing (as received). The value of the critical strain $_c$ at which the logarithmic decrement begins to increase with strain amplitude depends on annealing temperature and it depends on the kind of particles.

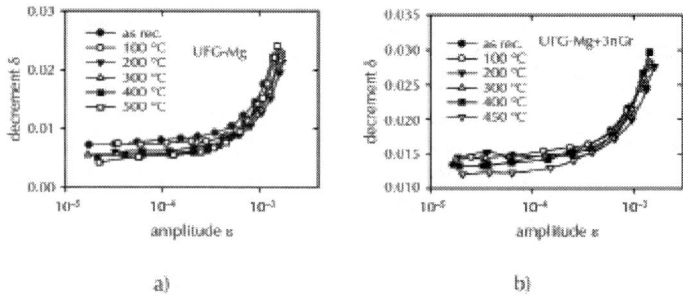

a)   b)

Figure 25: Strain amplitude dependence of the decrement for UFG-Mg (a) and UFG-Mg+3nGr (b), annealed at increasing temperature.

It is generally accepted that the $_0$ component is connected with the material microstructure and the amplitude dependent component $_H$ is due to an interaction of dislocations with point defects (e.g. solute atoms). The dislocations contribute to the damping capacity by the motion of vibrating dislocation lines. Particulates (Al_2O_3, ZrO_2, Gr), as secondary phases, may improve the damping capacity of alloys. The particles in a composite may cause an increase in the dislocation density as a result of thermal strain mismatch between the ceramic particles and the matrix during preparation and/or thermal treatment. The difference between the coefficients of thermal expansion (CTE) of the particles and the matrix may create the thermal residual stresses after cooling from the processing temperature to room temperature. The thermal stresses are also generated during heat treatment and when composites are heated and cooled through a temperature range (thermal cycling). In generally, there are biaxial or triaxial stress fields located near the reinforcement-matrix interface. The amplitudes of the stress field decrease with increasing distance from the interface. In a simple one-dimensional model, the thermal stresses $_{TS}$, produced by a temperature change T, at the interface are given by (Chawla, 1991)

$$\sigma_{TS} = \frac{E_F E_M}{\left(E_F f + E_M \left(1-f\right)\right)} f \Delta \alpha \Delta T \tag{9}$$

where   is the mismatch of the CTE values between the reinforcements and the matrix, E_F and E_M are the values of the Young's modulus for the reinforcement and the matrix, respectively and f is the volume fraction of the reinforcement. The CTE of the matrix is much higher than that of the reinforcement. Thus after cooling, the tensile thermal residual stresses are created in the matrix. The thermal stresses may relax around the matrix-reinforcement interface by emitting dislocation. An increase in the dislocation density near reinforcement has been calculated as (Arsenault & Shi, 1986, Dunand & Mortensen, 1991):

$$\Delta \rho = \frac{Bf \Delta \alpha \Delta T}{b\left(1-f\right)} \frac{1}{t} \tag{10}$$

where t is the minimum dimension of reinforcement, b is the magnitude of the Burgers vector of dislocations and B is a geometrical constant (depending on the aspect ratio). The newly formed dislocations may be sources of the higher damping capacity because of the motion of vibrating dislocation lines under cycling loading. According to the Granato-Lücke theory (Granato, & Lücke, 1981), weak and strong pinning points restrict motion of dislocations in their glide plane. At low strain amplitudes (stresses), dislocation bows out between the weak pinning points, leading to the amplitude independent logarithmic decrement. The dislocation remains anchored at the strong pinning points. At higher strain amplitudes, the force exerted on the weak pinning points becomes higher than the binding force of the weak pinning points. The dislocation segments break away from some weak pinning points within longer dislocation segments. This leads to a drastic instantaneous increase in the dislocation strain and thus giving rise to the level of the logarithmic decrement. The amplitude dependent component of the decrement depends on the dislocation density, length of shorter and longer dislocation segments. At temperatures higher than 0 K, the breakaway of shorter dislocation segments is thermally activated. During thermal cycling, new dislocations are created due to the difference in the CTEs (see Eq. (10)). An increase in the dislocation density, while the number of weak pinning points remains constant, increases effectively the length of dislocation segment between weak pinning points and hence the $_H$ component should increase, which is observed. The higher temperature of the cycling, the higher the dislocation density and the higher increase in the effective length of the shorter dislocation segments. The critical strain $_c$ at which the decrement becomes to be amplitude dependent is determined by the interaction energy between dislocations and the weak pinning points and the distance between the pinning points. As the distance between weak pinning point increases with the upper temperature, the dislocations can break away from the weak pinning point under lower critical stress and thus, the value of $_c$ should decrease with increasing upper temperature, which is observed. The values of $_c$ for a composite are proportional to the yield stress of the composite matrix. It should be consider that the value of $_c$ can be influenced by the thermal stresses due to the difference in CTEs for both components of the composite. In mc and UFG Mg no new dislocations are created during thermal cycling. Hence, it is expected only a slight influence of the thermal cycling on the versus curves, which is in agreement with experimental results.

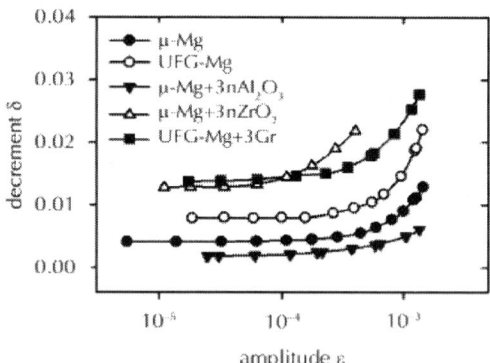

Figure 26: Amplitude dependence of the decrement for µ-Mg, UFG-Mg and µ-Mg and UFG-Mg reinforced with np.

Interfaces may contribute to the damping behaviour of composite materials. Due to a large difference in the coefficients of thermal expansion between the particles and the matrix, high thermal residual stresses are created around the particles. The thermal stresses with temperature can relax and plastic zones of tangled dislocation are formed. Their average radius depends on the difference in CTEs, particles radius and the change of temperature. The higher the matrix yield stress, the smaller radius of the plastic zones (Dunand & Mortensen, 1991). Plastic zones may also contribute to the damping behaviour.

The contribution of interfaces to internal friction depends on the character of interfacial bonding. The interface damping should be considered. The internal friction of a composite with the perfectly bonded interface (interface is incoherent) depends on the shape of particles, their volume fraction and the magnitude of local stress at the particle-matrix interface. Assuming that all the particles have the same diameter, the logarithmic decrement is given as (Zhang et al., 1994a)

$$\delta = \frac{4.5(1-v)}{\pi(2-v)}V_p$$

(11)

where V_p is the volume fraction of particles and   is the Poisson ratio of the matrix. The effect of the perfectly bonded interfaces on the damping behaviour becomes more significant at high temperatures; in Mg based composites this could be above 150 °C. The matrix becomes relatively soft relative to the nanoparticles. Hence, the internal friction at the interface may be thermally activated.

In the case of weak bonding at the interface, interfacial slip (sliding at the interface) may occur. In this case, the frictional energy loss caused by the sliding at interfaces may become a primary source of damping. The damping component due to interfacial slip under the applied stress amplitude $_0$ can be expressed as (Zhang et al., 1994a)

$$\delta = \frac{3\pi^2}{2}\frac{\kappa\sigma_r(\varepsilon_0 - \varepsilon_{cr})}{\sigma_0^2 / E_c}V_p$$

(12)

where   is the friction coefficient between both components of the composite, $_r$ is the radial stress at the particle-matrix interface corresponding to the stress amplitude $_{0'}$ $_0$ is the strain amplitude corresponding to $_0$. The critical interface strain $_{cr}$ corresponding to the critical interface shear stress $_{cr}$ is the strain at which the sliding on the on interface begins. E_C is the elastic modulus of the composite. For weakly bonded interfaces, the critical strain $_{cr}$ is assumed to be much lower than the strain amplitude $_0$ and then

$$\delta = \frac{3\pi^2}{2} \frac{\kappa \sigma_r}{\sigma_0} V_p$$

(13)

The stress concentration factor $/_0$ has been reported to be 1.1-1.3 (Zhang et al., 1994b). The model does not take into account possible effects of temperature or frequency on damping and therefore the predictions of the model may be taken only as the first approximation. Very high values of $_0$ for mc and nc Mg reinforced by zirconia np and Gr np are very probably caused by interfacial slip due to weak bonding between particles and matrix. The effect of an additional damping due to the influence of particle can be clearly seen in Figure 26. Therefore, the effect of an additional damping due to the influence of particles may be estimated. The measured increase in $_0$ due to the addition of ZrO_2 np to μ-Mg and Gr np to UFG-Mg is about 0.008-0.009 (Figure 26). Taking for the friction coefficient a typical value of ~0.08, the relation (10) predicts a value for an additional damping owing to interfaces of $_{0i}$ = 0.04. The discrepancy between the predicted and measured value may be caused by a non-uniform strain state in specimens because they were subjected to bending in experiment. Hence, the stress (strain) level can reach its critical value at which the interface sliding starts only in some region of the measured specimen. Accordingly, the measured interface-damping component is lower than the predicted value. High values of the decrement component $_0$ obtained for mc and UFG-Mg are very probably caused by grain boundary sliding support by diffusion processes. On the other hand, the observed decrease in the amplitude independent component of μ-Mg+3nAl_2O_3 comparing with μ-Mg may be explained according to the following way: Np situated in grain boundaries prevent the grain boundary sliding which contribute significantly to the amplitude independent component of decrement.

Temperature Relaxation Spectrum of Internal Friction

The temperature dependent internal friction (TDIF), internal friction spectrum, is the temperature or frequency dependence of damping. Internal friction spectra were measured in a DMA 2980 (TA Instruments) apparatus in a single heating and cooling process. Measurements were performed at various frequencies from 0.1 up to 2 Hz. Throughout the measurements the strain amplitude was 1.2×10^{-4}. The heating as well as cooling rate was 1 K/min.

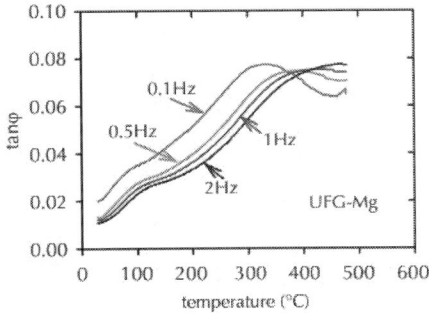

Figure 27: Temperature relaxation spectrum of IF estimated for various frequencies.

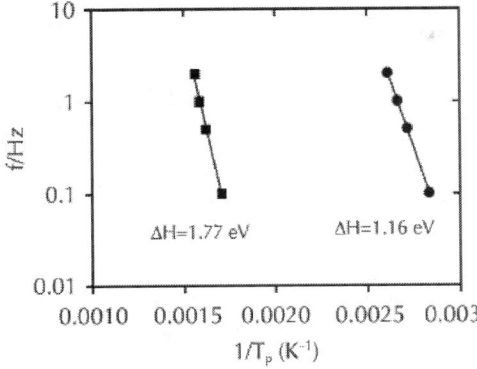

Figure 28: Arrhenius plots estimated for both peaks.

The internal friction in a material is a result of internal processes, which occurs during alternating stress cycles imposed on it and these processes originate from the interactions among structural components in the material. The presence of point, line, and planar defects within a stressed material often causes the internal friction to occur because of the atomic movement, rearrangement, or realignment of the defects under the application of a stress. These techniques can be used to characterize types of structure defects.

Figure 27 shows the temperature dependence of the internal friction IF=tan (Nowick & Berry. 1972) measured at 0.1 Hz during heating (heating rate 1 K/min.). Two peaks showing different intensities have been found: the weak low temperature peak (P1) and the more intense high temperature peak (P2). It can be seen that both peaks shift to a higher temperatures with increasing frequency, indicating that they may be relaxations peaks.

Table 7: Temperatures of the relaxation peaks estimated for various frequencies

f (Hz)	**P1 (°C)**	**P2 (°C)**
0.1	79.6	313.2
0.5	94.7	344.5
1	103.3	357.9
2	109.5	367.2

The curves introduced in Fig. 27 were measured during one heating course in the multi-frequent mode. The internal friction peaks are assumed to be imposed by a background IF_b expressed by

$$IF_b = A + B\exp\left(-C_b / kT\right) \tag{14}$$

where k is the Boltzmann constant, T is the absolute temperature, A, B, and C_b are constants. After subtracting the background by using a fitting program PeakFit, the peak temperature T_p was estimated for the both relaxation peaks. The temperatures of both peaks estimated for various frequencies are given in Table 7. The peak widths are broader than that for a Debye peak, characterised by a single relaxation time.

The internal friction peak appears at the condition $\omega\tau = 1$ (Nowick & Berry. 1972), with

$$\omega t = \omega t_0 \exp\left(DH / kT\right) \qquad (15)$$

where ω is the angular frequency (= $2\pi f$, f is the measuring frequency), τ is the mean relaxation time, τ_0 is the pre-exponential factor, and H is the mean activation enthalpy.

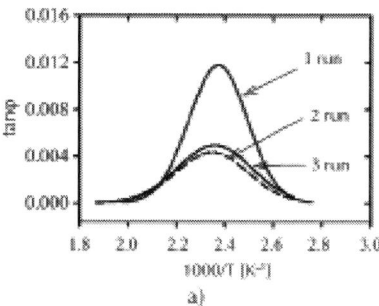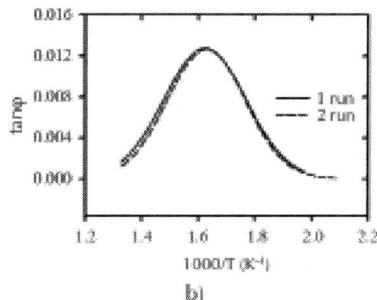

Figure 29: Background subtracted low (a) and high (b) temperature peaks after a prestraining of 2.55%.

Low Temperature Peak

Figure 28 shows the semilogarithmic plot of the frequency versus reciprocal value of peaks temperatures T_p (estimated for both peaks) so called Arrhenius plots. From the slope and intercept of the straight line, the mean activation enthalpy for the low temperature peak H has been obtained and its value is 1.16±0.05 eV.

As far as the low temperature peak may have a dislocation nature, prestraining of the sample at low temperature should influence the height of the peak. Therefore, the samples were deformed by rolling. Deformation by rolling at room temperature ($\varepsilon \approx 2.55\%$) increases the strength of the low temperature peak, as it is obvious from **Figure 29a** (the peak was extracted from the internal friction spectrum subtracting the background damping). However, the peak is stronger only in the first run after prestraining. In the second and third run, the height of the peak is approximately the same as in the as-prepared state, and the height is lower than in the first run.

The strain amplitude dependent component of the logarithmic decrement (Figure 25a) indicates free dislocation loops in the material. IF peaks estimated in the second and third heating and cooling runs are practically the same. Beside this original dislocation population, the prestraining of the sample produced new dislocations which were practically completely recovered during the heating course of measurements. Owing to small grain size the grain boundaries were very probably the sources as well as sinks of the newly created dislocations. As mentioned above, the dominant deformation mode in magnesium is the basal slip; glide of <a> dislocations. The secondary conservative slip may be realised by motion of <a> dislocations on prismatic and pyramidal planes of the first kind. Couret & Caillard (1984a,b) studied by TEM prismatic glide in magnesium over a wide temperature range. They have reported that the screw dislocations with the Burgers vector of $1/3[11\bar{2}0]$ are able to glide on prismatic planes and their mobility is much lower than the mobility of edge dislocations. The deformation is controlled by the thermally activated glide of those screw dislocation segments. A single controlling mechanism has been identified as the Friedel-Escaig mechanism. This mechanism assumes a dissociated dislocation on a compact plane (0001) that joints together along a critical length L_r producing double kinks on non-compact plane. Dislocations in very small grains are pinned at the grain boundaries and this is the reason why this dislocation population is so stable. On the other hand, the newly created dislocations during cold rolling are only slightly pinned and they are recoverable during heating part of the measurement cycle.

High Temperature Peak

The high temperature peak was observed at a temperature of ≈ 345 °C (0.5Hz). The cold prestraining of the sample did not affect the strength of the high temperature peak, as it can be seen from Fig. 29b. Position and height of the peak are very stable during heating as well as cooling. The activation energy was obtained from the frequency dependence of the peak temperature (Arrhenius plot) to be 1.77 ± 0.05 eV (see Figure 28).

The presence of the high temperature peak in magnesium has been reported by Kê (Kê, 1947). He described this peak as being related to the grain boundary relaxation.

Grain boundaries in materials with the small grain size contain a dense network of overlapping grain boundary dislocations. Grain boundary sliding is realized by the slip and climb (providing a maintenance of vacancy sources and sinks) of the grain boundary dislocations. Since both modes of the grain boundary dislocation motion must occur simultaneously, the slower one will control the grain boundary sliding. The climb mode involves jog formation and grain boundary diffusion and both modes of the grain boundary dislocation movement may be affected by the interaction of the grain boundary dislocations with impurities segregated in grain boundaries. According to Lakki et al. (1998) the activation energy for such process can be expressed as $H = H_j + H_{GB}$, where H_j and H_{GB} are the activation enthalpies for the jog formation and the grain boundary diffusion, respectively. The activation enthalpy for grain boundary diffusion in the coarse grained magnesium is 0.95 eV (Frost & Ashby, 1982). Comparing with the estimated activation enthalpy value of 1.77 eV, the possible jog formation energy should be approximately 0.8 eV, which seems to be a reasonable value. Taking into consideration that the surface of grains was passivated during the milling by 1% of oxygen in a protective argon atmosphere; therefore, MgO particles in the grain boundaries may influence the grain boundary diffusion characteristics.

ACKNOWLEDGEMENTS

Z.T. and P.L. are grateful for the financial support to the Czech Grant Agency under the contract P108/12/J018.

REFERENCES

1. Andersson, P., Cáceres, C.H. & Koike, J. (2003) Hall-Petch parameters for tension and compression in cast Mg. Mater. Sci. Forum 419–42, 123–128. ISSN: 1662-9752

2. Chawla, K.K. (1991) Metal matrix composites. VCH Weinheim, ISBN: 03872-85679

3. Arsenault, R.J. & Shi, N. (1986). Dislocation generation due to differences between the coefficients of thermal expansion. Mater. Sci. Eng., 81, 175-187. ISSN: 0254-0584.

4. Choi, H.J., Kim, Y., Shin, J.H. & Bae, D.H. (2010) Deformation behavior of magnesium in the grain size spectrum from nano- to micrometer. Mater. Sci. Eng. A 527, 1565–1570. ISSN: 0254-0584.

5. Couret, A. & Caillard, D. (1985a) An in situ study of prismatic glide in magnesium—I. The rate controlling mechanism. Acta Metall. 33 1447-1454. ISSN: 0921-5093

6. Courret, A. & Caillard, D. (1985b) An in situ study of prismatic glide in magnesium—II. Microscopic activation parameters, Acta Metall. 33, 1455-1462. ISSN: 0921-5093

7. Dieringa, H. (2011) Properties of magnesium alloys reinforced with nanoparticles and carbon nanotubes: a review. J. Mater. Sci. 46, 289–306. ISSN: 0022-2461

8. Dragomir, I.C. & Ungár, T. (2002) The contrast factors of dislocations in the hexagonal crystal system. J. Appl. Cryst. 35, 556-564.

9. Dunand, D.C. & Mortensen, A. (1991) On plastic relaxation of thermal stresses in reinforced metals. Acta Metall. Mater. 39, 127-139. ISSN: 0921-5093

10. Estrin, Y. & Vinogradov A. (2013) Acta Mater. 61, 782-817. ISSN: 0921-5093

11. Ferkel, H. (2003) Properties of graphite-strengthened magnesium. Adv. Eng. Mater. 5, 886-889. ISSN: 1527-2648

12. Ferkel, H. & Mordike, B.L. (2001) Magnesium strengthened by SiC nanoparticles. Mater. Sci. Eng. A 298, 193–199. ISSN: 0254-0584

13. Frost, H.J. & Ashby, M. F., Deformation-mechanism maps. Pergamon Press, Oxford, UK, 1982, 44. ISBN 0-08-029337-9

14. Granato, A.V. & Lücke, K. (1981) Temperature dependence of amplitude-dependent dislocation damping` J. Appl. Phys., 52, 7136-7142. ISSN 0021-8979

15. Ion, S.E., Humphrey, F.J. & White, S.H. (1982) Dynamic recrystallization and the development of microstructure during

hot deformation of magnesium. Acta Met. 30, 1909-1919. ISSN: 0921-5093

16. Kê, T.S. (1947) Experimental evidence of the viscous behaviour of grain boundaries in metals Phys. Rev. 71, 533-546. 1050-2947

17. Koike, T., Kobayashi, T., Mukai, T., Watanabe, T.H., Suzuki, M., Maruyama, K. & Higashi, K. (2003) The activity of non-basal slip systems and dynamic recovery at room temperature in fine-grained AZ31B magnesium alloys. Acta Mater. 51, 2055–2065. ISSN: 0921-5093

18. Kužel, R. & Klimanek, P. (1988) X ray diffraction line broadening due to dislocations in non-cubic materials. 2. The case of line anisotropy applied to hexagonal crystals. J. Appl.Cryst. 21, 363-368. ISSN: 0021-8898

19. Kužel, R. (1998) DIFPATAN—Program for Powder Pattern Analysis. http://www.xray.cz/ecm-cd/soft/xray/indexdifp.html

20. Lakki, A., Schaller, R., Carry, C. & Benoit, W. (1998) High temperature anelastic and viscoplastic deformation of fine-grained MgO-doped Al2O3`, Acta Mater., 46, 689-700. ISSN: 0921-5093

21. Langdon, T.G. (1994) A unified approach to grain boundary sliding and superplasticity. Acta Metall. Mater. 42, 2437-2443. ISSN: 0921-5093

22. Lukáč, P., Trojanová, Z., Száraz, Z., Svoboda, M. & Ferbl, H. (2006) Deformation behaviour of ultrafine grained magnesium with 3 vol. % graphite. Z. Metallkd. 97, 344-349. ISSS: 1862-5282

23. Mabuchi, M. & Higashi, K. (1999) On accommodation helper mechanisms for superplasticity in metal matrix composites. Acta Mater. 47, 1915-1922. ISSN: 0921-5093

24. Mabuchi, M., Chino, Y., Iwasaki, H., Aizawa, T. & Higashi, K. (2001) The grain size and texture dependence of tensile properties in extruded Mg–9Al–1Zn. Mater. Trans. 42, 1182–1189. ISSN: 1345-9678

25. Máthis, K., Nyilas, K., Axt, A., Dragomir-Cernatescu, I., Ungár, T. & Lukáč, P. (2004) The evolution of non-basal dislocations as a function of deformation temperature in pure magnesium

determined by X-ray diffraction, Acta Mater., 52, 2889–2894. ISSN: 0921-5093

26. Máthis, K., Čapek, J., Zdražilová, Z. & Trojanová, Z. (2011) Investigation of tension– compression asymmetry of magnesium by use of the acoustic emission technique. Mater. Sci. Eng. A 528, 5904–5907. ISSN: 0254-0584

27. Mohamed, F.A. (2003) A dislocation model for the minimum grain size obtainable by milling. Acta Mater. 51, 4107–4119. ISSN: 0921-5093

28. Moll, F., Chmelík, F., Lukáč, P., Mordike, B.L. & Kainer, K.U. (2000) Creep behaviour of a QE22-SiC particle reinforced composite investigated by acoustic emission and scanning electron microscopy. Mater. Sci. Eng. A 291, 246-249. ISSN: 0254-0584

29. Naser, J., Riehemann, W. & Ferkel, H. (1997) Dispersion hardening of metals by nanoscaled ceramic powders. Mater. Sci. Eng. A 234–236, 467–469. ISSN: 0254-0584

30. Nayeb-Hashemi, A.A., Clark, J.B., & Pelton A.D. The Li-Mg (Lithium-Magnesium) system, Bulletin of Alloy Phase Diagrams, 5, 365-374, 1984. ISSN: 1054-9714

31. Nes, E., Holmedal, B., Evangelista, E. & Marthinsen, K. (2005) Modelling grain boundary strengthening in ultra-fine aluminium alloys`, Mater. Sci. Eng. A, 410–411, 178–182. ISSN: 0254-0584

32. Nowick, A.S. & Berry, B.S. (1972) Anelastic Relaxations in Crystalline Solids Academic Press, New York/London. ISBN: 70-154378

33. Száraz, Z.. Trojanová, Z., Cabbibo, M., & Evangelista, E. (2007) Strengthening in a WE54 magnesium alloy containing SiC particles. Mater. Sci. Eng. A, 462, 225-229. ISSN: 0254-0584

34. Thostenson, E.T., Li, C. & Chou, T.-W. (2005) Nanocomposites in context. Compos. Sci. Technol. 65, 491–516. ISSN: 0266-3538

35. Trojanová, Z., Lukáč, P., Ferkel, H., Mordike, B.L. & Riehemann, W. (1997) Stability of microstructure in magnesium reinforced by nanoscaled alumina particles. Mater. Sci. Eng. A 234-236, 798-801. ISSN: 0254-0584

36. Trojanová, Z, Száraz, Z., Mielczarek, A, Lukáč, P. & Riehemann, W. (2008) Plastic and fatigue behaviour of ultrafine-grained

magnesium. Mater. Sci. Eng. A 483–484, 477-480. ISSN: 0254-0584

37. Trojanová, Z., Máthis, K., Lukáč, P., Németh, G. & Chmelík, F. (2011) Internal stress and thermally activated dislocation motion in an AZ63 magnesium alloy. Mater. Chem. Phys. 130, 1146-1150. ISSN: 0254-0584.

38. Ungár, T. & Borbely, A. (1996) The effect of dislocation contrast on X-ray line broadening: a new approach to line profile analysis. Appl. Phys. Let.; 69, 3173-3175. ISSN: 0003-6951

39. Wilkens, M. (1970) `The determination of density and distribution of dislocations indeformed single crystals from broadened X-ray diffraction profiles, phys. stat. sol. (a), 2, 359-370. ISSN: 1862-6319

40. Yoo, M.H., Agnew, S.R., Morris, J.R. & Ho, K.M. (2001) Non-basal slip systems in HCP metals and alloys: source mechanisms. Mater. Sci. Eng. A, 319-321, 87-92 ISSN: 0254-0584

41. Zhang, J., Perez, R.J., Wong, C.R. & Lavernia, E.J. (1994a) Effects of secondary phases on the damping behaviour of metals, alloys and metal matrix composites. Mater. Sci. Eng. R13, 325-390. ISSN: 0254-0584

42. Zhang, J., Perez, R.J. & Lavernia, E.J. (1994b) Effect of SiC and graphite particulates on the damping behavior of metal matrix composites. Acta Metal. Mater., 42, 395-409. ISSN: 0921-5093

10

Installation Effects of Auger Cast-in-place Piles

Fathi M. Abdrabbo, and Khaled E. Gaaver[*]

Structural Engineering Department, Faculty of Engineering, Alexandria University, Alexandria, Egypt

ABSTRACT

Since their introduction in Europe and North America some 50 years ago, auger cast-in-place piles (ACIP) have become increasingly popular all over the world. These piles offer considerable environmental advantages during construction including minimal vibration, and low noise beside their high productivity. The most severe limitation of the ACIP is its sensitivity to operator performance, which can lead to a pile of poor integrity or inconsistent quality. Thus the improper use of ACIP equipment can result in piles containing defects or can cause instability of nearby structures. Three case studies are presented and discussed in an effort to illustrate learned lessons. First case study highlights the misuse of ACIP equipment leading to unreliable defective pile foundations. Second and third case studies show the adverse effects of installing ACIP on the stability of nearby structures. The study revealed that it is essential to employ a clever pile crew during the installation of ACIP to observe, interpret, and take corrective actions for unusual situations. The authorities worldwide should oblige pile contractors to employ only experienced and qualified workers in charge of geotechnical engineering works. Tender documents should include precise clauses related to the technological factors affecting the quality of ACIP. Unfavorable side effects of installing ACIP in saturated loose and medium sand can cause tilt of adjacent existing structures; even they are on either shallow or deep foundations. A row of micro-piles and/or soil grouting adjacent to the existing buildings were successfully used to reduce the adverse effects of ACIP. Implementation of different codes on the results of pile loading tests produced different pile working loads. Therefore tender documents should specify the code upon which interpreting the pile test results. At the meantime the geotechnical engineer should implement his experience and judgment during application of the specified code. Finally this work indicates that the outcome prediction of ACIP may deviate from the actual performance.

INTRODUCTION

Auger cast-in-place piles (ACIP), also known as continuous flight auger (CFA) piles, are being increasingly used to support different structures around the world especially in the Middle East. ACIP are installed by means of an auger with a hollow stem having an inner diameter of 100–200 mm. The hollow stem is provided by a temporarily closure plate at the bottom. The auger is inserted into the soil by the combined action of an axial thrust and a torque. After reaching the desired depth, the auger is pulled up 300 mm and then the closure plate is pushed out by pumping a high slump concrete through the stem. The auger is lowered to its original depth and then it is withdrawn at a controlled rate while pumping concrete through the stem. The auger withdrawn allows removing the soil retained on auger flights and forming a shaft of concrete extending to ground level. Reinforcing cage with suitable centralizing spacers attached can be installed by pushing into the wet concrete. The procedure allows the installation of piles with a diameter of 400–1000 mm and a length up to 35 m. But the length of steel cage is limited to about 20.00 m. ACIP has the advantages of suitability for most soil formations, rapid construction, environmental friendliness, less labor, and easy-machinery maintenance. On the other hand, ACIP has some disadvantages due to the implementation and the misuse of the pile equipment. The behavior of ACIP is intermediate between bored and driven piles and strongly influenced by the installation procedures [15].

Many uncertainties may be involved in the construction of ACIP. These uncertainties lead to inappropriate construction process and increase of the construction cost due to the remedy procedure of the defected piles. The sources of uncertainties in geotechnical investigations were reported by Abdrabbo [1]. The imperfections in ACIP resulting from the construction process were explored by Poulos [19], and Abdrabbo and Abouseeda [2]. Uncertainties in the interpretation of pile loading tests were reported by O'Neill and Hassen [17], and Poulos [18]. It is difficult to define and set a database for all uncertainties arising during the construction of ACIP due to dissimilar machinery implemented in construction, different types of soils, and uneven skills of workers all over the world. In addition, most geotechnical engineers pay their attention to the installed piles in terms of drilling, pumping concrete, and insertion of steel cage irrespective of side effects resulted from the technological factors involved in the used machinery. The soil around the installed piles suffers tremendously disturbance from the imposed stresses resulting from drilling process and pumping concrete, and thus the soil properties may be changed. If the volume of the removed soil is less than the theoretical volume of the pile, the net resulting effect is a compression of the soil surrounding the pile. Neely [16] expressed the soil disturbance due to the construction of ACIP as soil decompression. The term soil decompression refers to the reduction in soil stiffness and strength brought about by the drilling action of the continuous flight auger. Fortunately, the recovery of soil from such a disturbance may not require a long time, Lee and Poulos [14].

Neely [16] proposed a relationship between the rotational speed of the auger and the advance speed with the pitch of the flight auger to control the soil decompression while drilling a piled hole. Mandolini et al. [15] reported an expression to relate the diameter of the formed pile with the rate of concrete pumped through the hollow stem of the auger and the auger retrieval rate. Viggiani [22] defined the critical rate of the penetration of the auger into the soil as a function of the rate of revolution, the overall diameter of the auger, the outer diameter of the central hollow stem, and the pitch of the auger. Vipulanandan et al. [23] showed that the critical rate of penetration of the auger is about 30 mm/s with an auger revolution of 5 rpm. If the advance velocity is greater than the critical velocity, compression effect in soil surrounding the pile hole occurs. On the other hand, decompression in the soil surrounding the pile hole occurs if the advance velocity is less than the critical velocity. Moreover, Viggiani [22] showed that a minimum value of axial thrust is required to ensure that the advance velocity of the auger is equal to or greater than its critical velocity. It is noticeable that both the expression reported by Viggiani [22] and the condition given by Vipulanandan et al. [23] are independent of the soil type and properties. These expressions need to be justified.

The effect of grout ratio on the performance of ACIP was investigated by Neely [16], and Abdrabbo and Mahmoud [4]. Neely [16] pointed out that as the grout ratio decreases the grout placement pressure decreases which may lead to contamination of the grout/concrete with soil. Abdrabbo and Mahmoud [4] reported that as the grout ratio of ACIP in sand increases from 1.20 to 1.60, the ultimate base load of the pile increased by 24% without noticeable effect on the shaft load. Abdrabbo and Gaaver [3] termed the soil disturbance as excess pore water pressure resulting from cyclic shear stresses on the adjacent soil. Their study outlined the geometry of the influenced zone due to drilling a hole by an auger.

Discussion and analysis of construction observations is a crucial aspect in geotechnical engineering field. This technique was termed as the observational procedure and reported by many authors such as Wu [24] and Glass and Powderham [11]. Through observational procedure, this study aims to define some sources of deficiency in ACIP to improve their technology. Moreover, the current study objects to share the practice gained from three case studies with the geotechnical engineering community worldwide and to add this practical knowledge to the engineering literature.

CASE STUDY NO. 1

Site Description and Subsoil Formations

The foundations of three typical educational buildings and two service buildings were constructed using ACIP at Smouha district, Alexandria, Egypt. The educational buildings consist of a ground floor in addition to six typical floors, while the service buildings are composed a ground floor and three typical floors, as shown in Fig. 1. All of these buildings were constructed using reinforced concrete skeletons with in filled brick walls. Prior to construction, 45 boreholes were drilled at the site up to a depth of 50.00 m below the ground surface to explore the subsoil conditions. The recovered soil samples were classified in accordance with ASTM D 2487. Geotechnical explorations revealed that the successions of soil layers are relatively uniform all over the site. Fig. 2 presents a typical succession of soil strata at one borehole.

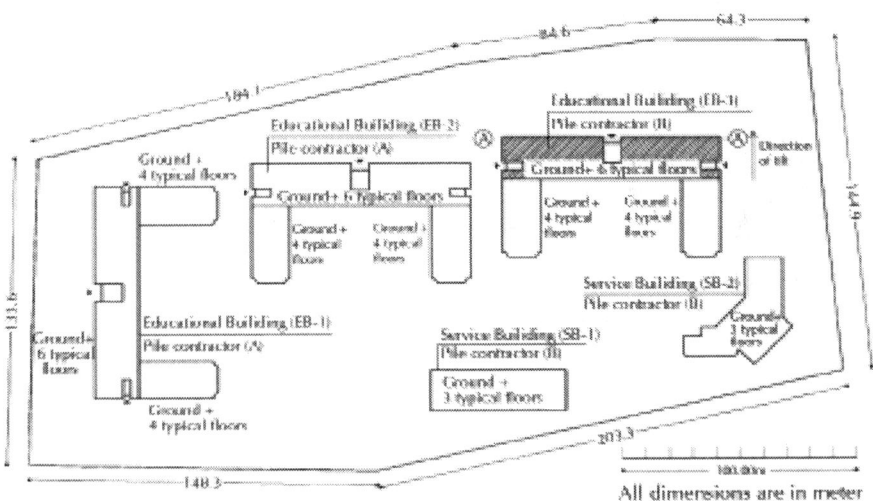

Figure 1: Layout of the buildings in case study #1.

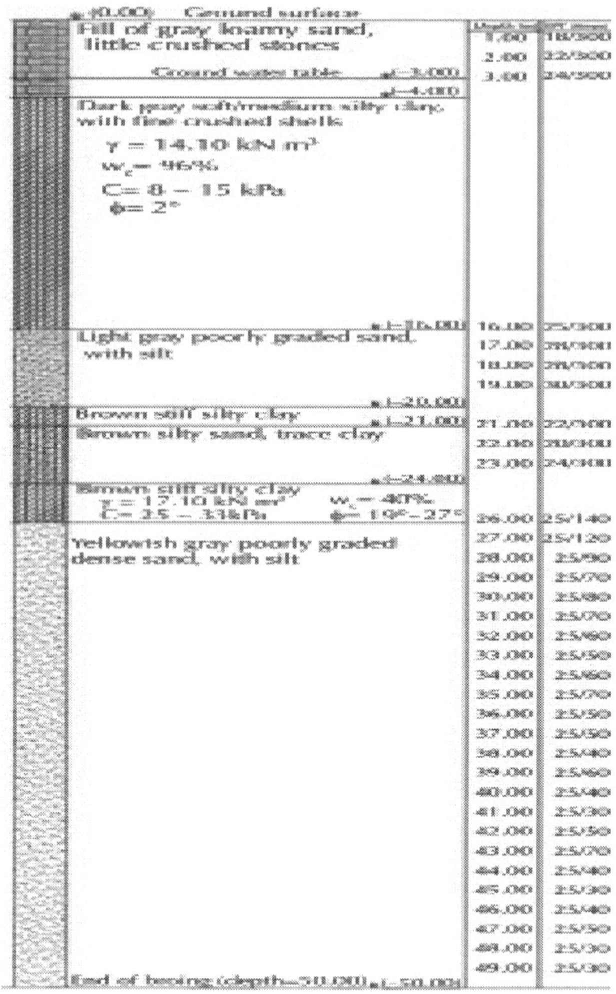

Figure 2: Typical borehole in case study #1.

The subsoil formation at the site consists of a top fill layer of loamy sand containing little crushed stones and extending to a depth varying from 4.00 to 5.00 m below the ground surface. Notably, the top fill layer was recently placed because the site was a swamp area of low level. The fill layer is underlain by soft silty clay, which extends to a depth of 16.00 m below the ground surface. A bed of poorly graded dense sand was encountered at a depth of 16.00 m and extended up to the end of exploration. The sand is intervened by a layer of stiff silty clay having different thicknesses. The top surface of the stiff clay was encountered at depth varied from 20.00 to 22.00 m below the ground surface. In most boreholes, the top surface of stiff clay is at a depth 20.00 m and extended to different depths up to a depth of 27.00 m below the ground surface. It is recognized that the existence of stiff clay within the sand bed caused difficulties in assessing the appropriate pile length since the thickness of stiff clay varied from 2.00 to 7.00 m. The groundwater table was measured and found to be at 3.00 m below the ground surface.

The results of standard penetration tests at depths below 26.00 m indicated very high values. These values should be interpreted with caution due to the many factors affecting the test results. The undrained shear strengths of cohesive soils were measured using direct shear box apparatus on undisturbed clay samples. The test results conducted on representative soil samples are illustrated in Fig. 2.

The Piles

ACIP are recommended to support the superstructure loads. The piles extend to 31.00 m below the ground surface. Piles with diameters of either 500 mm or 600 mm are recommended. The reasons for adopting this type of pile are the limited construction time, lower cost, and the availability of many pile rigs. The working pile load was calculated in accordance with Egyptian code [10]. Down drag load on the piles resulted from recent fill placed on the ground surface before installing the piles was considered in the calculations of pile load. The down drag load is developed along the piles from the ground surface up to the neutral plane due to down movement of soft clay relative to the piles. The neutral plane was assumed to be at the lower surface of the soft clay layer. The developed down drag load and the down drag are time dependent; they increased as the time elapsed. Poulos and Davis [20] stated that in groups of end-bearing piles, the down drag loads on piles in a group are smaller than on an isolated pile. Lee and Ng [13] reported that the reduction in the down drag on a pile within a pile group with respect to single pile is sensitive to the relative stiffness between the consolidated clay and the end bearing stratum. They concluded that the shielding effect of down drag becomes larger as the stiffness of the bearing stratum increases. However Lee and Ng [13] and Chow et al. [7] showed that the shielding effect of down drag load on center pile of an end bearing pile group is insensitive to the relative stiffness between the consolidated clay and the end bearing stratum. The -method [6] was used for calculating the down drag load. The pile groups under the educational buildings were arranged in symmetrical pattern of two and four piles. Consequently the down drag load on a pile group was considered as the summation of down drag loads on individual piles in the group. Therefore the working loads of piles are 0.90 and 1.30 MN for the 500 and 600 mm diameter piles respectively.

The Problem

Value engineering concept was implemented in the design stage in a way that the engineer used mix of pile groups containing 500 and 600 mm pile diameters in the pile arrangement under the buildings. However, piles of different diameters have not been used in the same pile group. The distribution of pile groups and piles in the groups are exactly the same under the three educational buildings. Therefore, it was anticipated that the three educational buildings would behave similarly as long as the buildings and the subsoil conditions under them are identical. Despite these similarities, one of the three educational buildings, EB-3, tilted once the construction was completed and the buildings entered into service. After 3 years in service, the building EB-3 was put under study and observations.

Precise monitoring of the building EB-3 using a total station of accuracy ±1 mm over 24 months revealed that the average horizontal displacement at the uppermost point is 5.00 mm per month. Moreover, the rate of the lateral displacement was constant over the monitoring period. The total lateral displacement of the tilted building was back calculated to be 300 mm for a lifetime of 5 years. The total value of the lateral displacement was confirmed by measuring the gap formed at the expansion joint of the building. The width of the footprint of the building foundations is about 15.20 m and its height is 21.00 m, so the calculated differential settlement within the lifetime of 5 years is 217 mm. The lateral displacement and the differential settlement of the building are unacceptable according to the international codes. Therefore, it is essential to investigate the causes of the building tilt and the reason that one building tilted while the others are worked perfectly. These inspections include the study of the stability of fresh concrete during filling the pile holes, confirmation of pre-construction geotechnical investigations by drilling additional boreholes at the tilted side of the building, inspecting the daily reports during the pile installation, and studying the results of pile loading tests.

Inspections and Site Observations

To study the stability of fresh concrete in the pile hole through the soft silty clay layer, the equilibrium equation developed for stability of concrete inside an excavated slurry trench was implemented. This equation was developed for the stability of concrete in a slurry trench and not for circular piles, and some differences in earth pressures are expected. The implementation of this stability equation is not taking into consideration the effects of nearby piles, which previously constructed. The pre-constructed piles reinforce the soft layer and produce a composite geomaterial, which may change the value of the coefficient of lateral earth pressure and the shear strength of soil. The coefficient of earth pressure at rest for normally consolidated clay was predicted according to the plasticity index from Alpan [5] and Holtz and Kovacs [12]. The two equations give the value of (ko) equal to 0.60 and 0.70 respectively. An average value of 0.65 was considered in the analysis. The stability of fresh concrete was checked using Xanthakos [25] as:

$$Cu \geq (\gamma c \cdot kc - \gamma s \cdot ko) \cdot H / Nc \qquad (1)$$

where c and s effective unit weights of concrete and soil respectively, kc and ko coefficients of lateral pressure for concrete and soil respectively, H depth below ground surface, Nc bearing capacity factor.

The above equation is applied at the bottom surface of silty clay layer, 16.00 m below the ground surface, using c = 12 kN/m³, kc = 0.70, s = 4.10 kN/m³, ko = 0.65, and Nc = 9.0. This equation shows that the undrained shear strength of soft silty clay should be greater than 10.20 kPa to ensure the stability of fresh concrete while pumping in the pile hole. Clearly, the estimated shear strength is greater than the lower measured value by 27.5% and smaller than the upper measured value by 32%. Thus, the lateral collapse of soft clay may occur while pumping fresh concrete in the pile holes.

In such soil conditions, some precautions should be considered during pile construction such as filling the drilled hole with fresh concrete and re-drilling the hole in the same place through the fresh concrete. After the second stage of drilling, the concrete is pumped in, and a steel cage is inserted in place to complete the pile. This procedure is called an ACIP doublex. As an alternative procedure, the pile hole can be filled with cement-bentonite slurry. After setting the slurry, the pile hole can be re-drilled and filled by concrete, and steel cage is inserted to finish the pile. In our case study, the daily field reports indicated that none of these precautions were considered during the installation of the piles.

To verify the pre-construction soil investigations, two additional boreholes of 50.00 m in depth were conducted at the side of the building which exhibited excessive displacement. The soil samples obtained from new boreholes confirmed the soil formations and properties as indicated in the pre-construction geotechnical explorations. Another target of the additional boreholes is to confirm the continuation of the sand bed up to 50.00 m depth, without any interbedded clay pocket. Retrieving of soil samples from boreholes for visual inspection is preferable than interpreting of soil types from sounding field tests. Therefore, it is revealed that the soil conditions are not the cause of the tilting of the building. Another reason should be explored.

The daily field reports of the piles during the construction in the whole project were thoroughly investigated. It was discovered that the piles were constructed by two different contractors, contractor (A) and contractor (B), as shown in Fig. 1. The contracts for the two educational buildings (EB-1, and EB-2) were awarded to the first contractor (A), and the contracts for the third educational building (EB-3) and the two service buildings (SB-1, and SB-2) were awarded to the second contractor (B). It is valuable to reminder the reader that the tilted building is the third educational building (EB-3). Unfortunately, the technological factors such as rotational speed of the auger, advance speed of the auger, imposed torque were not recorded in the daily field reports. The reports indicated that the grout ratio of the installed piles at the site is about 1.20 for piles of 500 mm, whereas this ratio is approximately 1.14 for piles of 600 mm. The grout ratio is defined as the ratio of the actual volume of concrete to the theoretical volume of the pile. The minimum grout ratio suggested by the Deep Foundation Institute [9] is 1.15. This condition was satisfied for all of the piles of 500 mm in diameter, but it is critical for the piles of 600 mm in diameter.

During the installation of piles of the defective building (EB-3), the pile contractor executed 441 piles at the site in 27 working days using two drilling pile rigs. Thus, the level of soil disturbance due to drilling the pile holes at the site was very high. Two pile rigs with a pile crew on each rig were involved in the pile installation of (EB-3). It is expected that the operators of the two pile rigs have different experiences and will install piles of different qualities. In such situations, pile loading tests should be conducted on piles chosen from the piles installed by both pile crews to verify the working load of piles installed by each pile crew individually. The number of tests should not be less than two for each pile crew.

It was noted that contractor (B) provided pile rigs without Kelly bars, so the rig crew extended the auger using a short helical piece similar to the auger attached to the rotary head of the drilling machine. While drilling a pile hole, the drilling process started by the attached auger on the rig, and then drilling stopped to attach the short piece to the auger. If the flights of the auger are not match the flights of the short piece, interruption of the path of the excavated materials to ground surface occurs, and the excavated materials move laterally into adjacent soil. During the pumping of concrete while retrieving the auger, the retrieval operation was interrupted to take out the short piece. This process leads to hardening of the concrete in the hole and produces separation in the pile installed.

The quality of concrete used to form the piles is 30 N/mm², with maximum nominal size of gravel of 19 mm. The amount of water in concrete was just sufficient for cement hydration, where the water-cement ratio was 0.40. Additives were used to increase the workability of concrete. The rig crews of contractor (B) reported that subsidence in concrete in some of the pile holes occurred after complete filling with fresh concrete. An upward spring of water was also observed through the fresh concrete after pumping concrete in the pile hole. A subsidence of 0.40–2.40 m of the fresh concrete in some of the pile holes was recorded. The subsidence of fresh concrete in the pile holes and the water spring in concrete are due to the excess pore water pressure induced in the soil around the pile hole. The excess pore water pressure is due to cyclic shear stresses resulted from pile drilling.

This excess pore water pressure is accompanied by loss in the shear strength of soil [3]. Daily field reports of contractor (A) showed that none of these observations were recorded during the installation of piles in buildings (EB-1, and EB-2).

Twenty pile loading tests were conducted on randomly selected working piles at the site. Four loading tests for each building were planned. Ten tests were conducted on piles with a diameter of 500 mm while other ten tests were carried out on piles with a 600 mm diameter. The test loads were 1.50 and 2.10 MN for 500 and 600 mm respectively. The test loads are greater than 1.5 times the estimated pile working load to allow for down drag loads on the piles. Fig. 3 presents the load-displacement relationships of piles tested at the tilted building. Table 1 shows a summary of the test results for all piles tested in the project. The table depicts the measured displacements of piles tested at working loads and at test loads. The allowable displacement of the tested piles at 1.5 times the pile working load, according to Egyptian code [10], is calculated as 13.55 and 15.56 mm for 500 and 600 mm diameters, respectively. Table 1 confirms that the pile displacements at 1.5 times the working loads are acceptable for all of the tested piles except for pile No. 12 in building (EB-3) which indicates excessive displacement.

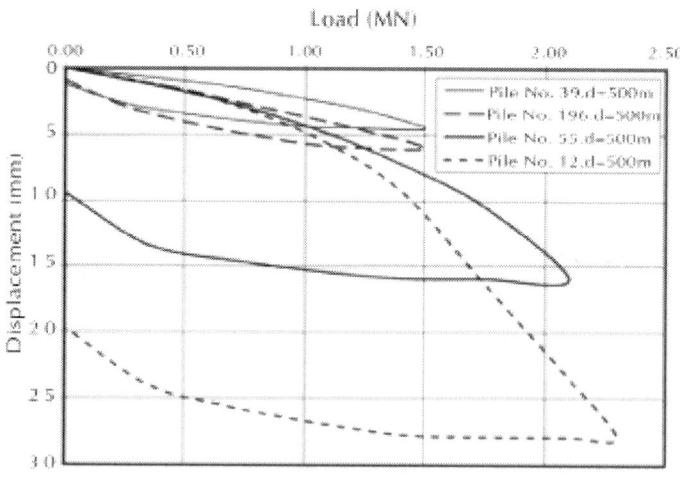

Figure 3: Load–displacement relationships of piles tested in the tilted building.

Table 1: Summary of pile loading tests

Pile no.	Pile diameter (mm)	Working load (P_{all}) (kN)	Displacement at working load (mm)	Displacement at 1.5 (P_{all}) (mm)	Allowable displacement at 1.5 (P_{all}), [10] (mm)	Building
15	500	900	2.82	4.60	13.55	EB-1
61	500	900	2.12	4.75	13.55	
224	600	1300	2.92	4.82	15.56	
307	600	1300	2.86	6.06	15.56	
19	500	900	2.00	4.31	13.55	EB-2
181	500	900	2.90	5.35	13.55	
315	600	1300	4.10	12.50	15.56	
411	600	1300	3.05	7.25	15.56	
39	500	900	2.00	3.50	13.55	EB-3
196	500	900	3.45	5.10	13.55	
55	600	1300	6.90	13.80	15.56	
12	600	1300	9.30	20.50	15.56	
12	500	900	3.02	5.80	13.55	SB-1
98	500	900	2.08	3.65	13.55	
81	600	1300	3.74	6.68	15.56	
189	600	1300	3.68	6.35	15.56	
18	500	900	3.05	5.95	13.55	SB-2
102	500	900	2.85	4.65	13.55	
91	600	1300	4.26	11.82	15.56	
195	600	1300	4.05	9.20	15.56	

The predicated ultimate loads of the tested piles were analyzed using Egyptian code [10] in addition to five European practices [8]: Czech Republic, French, German, Italian, and Norwegian. According to the criterion reported in each code, the predicated ultimate loads for piles of 500 mm in diameter varied from 1.60 to 3.80 MN, and these loads are approximately 1.78–4.22 times the design pile working load. For piles of 600 mm in diameter, the predicted ultimate pile loads are 2.20 and 4.52 MN, and these loads are approximately 1.69–3.48 times the design pile working load. Therefore, the factor of safety against soil–pile failure is acceptable. Table 2 illustrates a summary of the allowable pile load for all piles tested in the site. The variation in the values of the predicated working pile load is due to the different concepts and different factors of safety implemented in each method of analyzing the results of pile loading tests [8]. Therefore it is difficult to adopt a unique value for the pile working load. Tender documents should specify the code upon which conducting tests and interpreting of test results. At the meantime geotechnical engineer should implement his experience and judgment during the application of the specified code.

Table 2: Allowable pile load obtained from different codes in (kN)

Pile no.	Diameter (mm)	Czech	French	German	Italian	Norwegian	Egyptian	Building
15	500	1173	1250	1000	901	951	802	EB-1
61	500	1173	1250	1000	1225	1112	1449	
224	600	1642	1750	1400	1508	1454	1616	
307	600	1642	1750	1400	1524	1462	1648	
19	500	1172	1250	1000	1092	1046	1185	EB-2
181	500	1172	1250	1000	1124	1064	1258	
315	600	1642	1750	1400	1234	1252	1067	
411	600	1642	1750	1400	1523	1461	1645	
39	500	1172	1250	1000	1168	1084	1336	EB-3
196	500	1172	1250	1000	1499	1250	1602	
55	600	1642	1750	1400	1476	1354	1552	
12	600	1642	1750	1275	1237	1186	1075	
12	500	1172	1250	1000	910	953	821	SB-1
98	500	1172	1250	1000	1066	1025	1131	
81	600	1642	1750	1400	1364	1382	1327	
189	600	1642	1750	1380	1456	1340	1482	
18	500	1172	1250	1000	1126	1063	1252	SB-2
102	500	1172	1250	1000	1126	1063	1252	
91	600	1642	1750	1400	1293	1315	1186	
195	600	1642	1750	1400	1250	1325	1098	

Pile integrity tests using the echo method were conducted on random piles, and the test results were investigated. The test results revealed unreliable information about pile integrity due to the pile length, which may lead to unreliable reflected sound waves. Cross hole sonic tests are more reliable in the case of long piles. In addition, the quality of grouted concrete was observed using the results of cube crushing strength tests. The test results confirmed the concrete design strength.

The structural analysis of the building was reviewed without any sign of defects. According to these investigations, it seems from theories point of view that the piles installed at building (EB-3) should carry the imposed load from the building safely without any anticipated defects. Therefore, uneven settlement is not expected. Thus, the uneven lateral displacement of the building by 300 mm at the uppermost top level of the building is questionable.

To calculate the anticipated displacement of the pile groups, linear elastic theory and interaction factors developed by Poulos and Davis [20] were implemented. The pile groups supporting the building columns composed of 2, 3 and 4 piles. If we consider that the modulus of elasticity of concrete (Ep) and of very dense sand (Es) are 20,000 MPa and 80 MPa respectively, Ep/Es becomes 250. Therefore the interaction factors at pile spacing of 2.5 and 3.5 times the pile diameter are 0.35 and 0.30 respectively [20]. Thus, the displacement of four equally loaded piles in a group is 2 times the displacement of a single pile, whereas the displacement of two equally loaded piles in a group is 1.35 times the displacement of a single pile.

According to the worst results of loading tests, pile No. 12 (EB-3) which indicated a pile displacement of 9.30 mm at the working load, the expected settlement of the 4-pile group is 18.60 mm while the displacement of the 2-pile group is 12.56 mm. Therefore, the expected uneven settlement is 6.04 mm, which is incomparable to the measured value of 217 mm. The calculations are based on free-standing pile groups and the load carried by the soil underneath the pile caps is neglected. The calculations of the settlement of the building ignore the interaction between the piles in the adjacent groups and the piles in the considered group. Therefore if the piles of the defective building were installed properly, the building should not suffer any noticeable tilt.

The quality of ACIP is dependent upon many technological factors such as rotational speed of the auger, advance speed of the auger, auger pitch, weight of the auger attached to the rig, internal diameter of the hollow stem, imposed torque, grout pressure, grout ratio, and thrust load on the auger. The allowable pile load is depending upon these technological factors in addition to the soil conditions. It is expected that the abovementioned factors differ from one site to another and from one pile rig operator to another. This explains the reasons for the stability of the other two educational buildings because their piles were installed by another contractor. The tender documents of the project did not include any technical clauses relating to drilling equipments, drilling process, and associated machinery. The technical specifications of the contract were concentrated on the geotechnical aspects and pile specifications. Therefore tender documents should include precise clauses related to rotational speed of the auger, advance speed of the auger, auger pitch, imposed torque, grout pressure, grout ratio, and thrust load on the auger. The auger should be of height equal to the full length of the proposed piles, no extension to the auger should be used. It is essential to monitor these factors during installation of ACIP at the site. Automated system can be accommodated in the pile rig to monitor the rotational speed of the auger, advance speed of the auger, applied torque, grout pressure, thrust load, and volume of pumped concrete. The rig crew of piles should report daily to the engineer the geotechnical and technological aspects during pile installation. The recorded field observations during installation of piles should be analyzed and corrective actions should be immediately taken. It is important to emphasize that close and precise supervision of pile construction by a geotechnical engineer is necessary. Close communication between the engineer and the pile crew is significant.

Remedy Procedure

It was not wise to leave the building tilting, so a remedy procedure should be designed to keep the building in service. Two rows of micro-piles were designed and installed at the side of the building which exhibiting greater settlement, as shown in Fig. 4. Micro-piles of type (C) were used [21]. The diameter of the micro-piles is 200 mm, with spacing of 700 mm in longitudinal direction, whereas the spacing between the two rows is 1750 mm. The outside row of piles extends to a depth of 33.00 m below the road level, whereas the inside row of piles extends to a depth of 35.00 m below the ground floor level.

The difference in the pile length is due to the difference in the levels between the road and the ground floor. A cement grout with a ratio of 1: 1 (type I cement: water) was pressurized to grout the soil under the pile tips through a pipe of 128 mm inner diameter, 8 mm wall thickness, and provided with a nozzle at the tip. The pipe was also filled by the grout. A dispersing and self retarding agent was added to the cement grout in order to aid in efficient placement. Each micro-pile was completed by pressure injection to flush grout return to the ground surface along the micro-pile annulus. The grout ratio was recorded during the construction of the micro-piles. It varied between 1.25 and 2.40 while, in some piles, the grout ratio was 3.20 without any appearance of cement grout on the ground surface through the annular space between the steel pipe and the geomaterial.

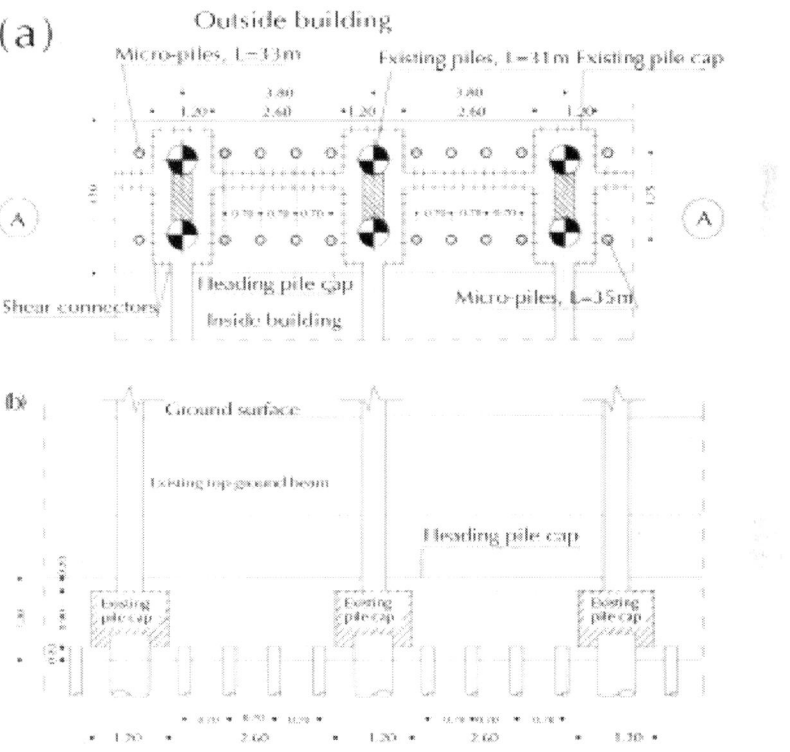

Figure 4: (a) Layout of micro-piles and existing piles, and (b) section elevation of pile caps.

The working load of micro-piles was calculated in accordance with the design concept reported by Schaefer [21]. A pre-construction pile loading test was conducted on single micro-pile to a test load of 750 kN. The pile settlement relationship is presented in Fig. 5. According to the test results, a pile working load of 250 kN was considered. Moreover four loading tests on four working micro-piles were conducted up to test loads of 400 kN, as shown in Fig. 5. The micro-piles were connected to the existing pile group through a heading pile cap. The heading pile cap was connected with the existing pile caps through shear connectors of steel dowels. During the excavation for the heading pile cap, it was observed that one of the existing piles was exhibiting a failure state at pile head just underneath the pile cap and that the concrete was contaminated with clay. Furthermore some of the pile dowels in the failed concrete portion were buckled, and some other dowels were short and not connected to the existing pile cap. It was noting that the defective pile is in a group of two piles carrying a corner column of the building. Thus the un-defective pile in this group may carry approximately double its working load; thus, the two piles in the group were almost in a failure state, and the nearby piles were overloaded. According to these observations, it is confirmed that the cause of the tilt of the building is the bad workmanship of the installed piles and improper engineering of the work. After remedy, the lateral displacement of the building became nil over 12 months.

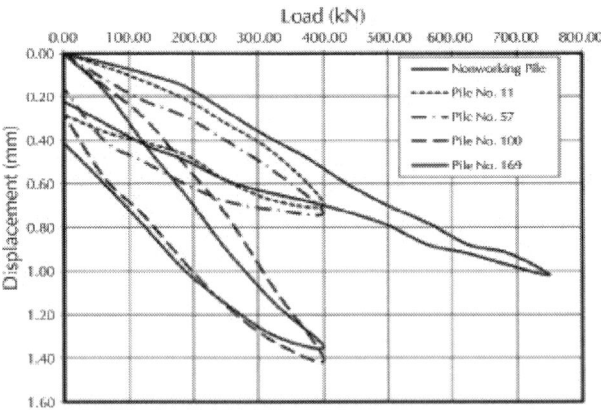

Figure 5: Load–displacement relationships of the micro-piles.

Learned Lessons

Based on this case study, it can be concluded that the choice of the pile contractor is vital in executing a proper job. The authorities worldwide should oblige the pile contractors to employ only experienced and qualified workers in charge of geotechnical engineering works. The workers and foreman should be licensed from official institutes. Implementation of different codes on the results of pile loading tests produced different pile working loads. Tender documents should specify the code upon which conducting tests and interpreting of test results. At the meantime the geotechnical engineer should implement his experience and judgment during the application of the specified code. Tender documents should include precise clauses related to rotational speed of the auger, advance speed of the auger, auger pitch, imposed torque, grout pressure, grout ratio, and thrust load on the auger. The auger should be of height equal to the full depth of the proposed piles, no extension to the auger should be used. It is not advisable to employ two pile rigs in a project unless pile loading tests on the production piles of each pile crew are conducted. Close and precise supervision of pile construction by a geotechnical engineer is essential. The rig crew of piles should report daily to the engineer the geotechnical and technological aspects during pile installation. The field observations on pile installation should be interpreted and reviewed. Corrective actions are immediately required. Sometimes the prediction of the functionality of ACIP may deviate from the actual performance.

CASE STUDY NO. 2

Site Description and Subsoil Formations

The second case study considers the adverse effects of cyclic shear resulting from the installation of ACIP on the properties of the surrounding soil and consequently on the stability of nearby structures. This problem attracted the attention of the writers two decades ago because of the legal consideration arising from the adverse effect of the installation of ACIP on the stability of adjacent buildings. In addition, the problem has drawn the attention of many authors worldwide such as Neely [16] and Mandolini et al. [15]. According to civil law, it is forbidden to influence the stability of any structure nearby or adjacent to the construction site. The civil law has no tolerance for these adverse effects. Thus, these effects are considered the major drawback of ACIP. The problem was illustrated through the following case study.

In a construction site located at Damanhur, West of the Nile River delta, El-Behaira province, Egypt, the site is adjacent to two residential buildings located at the western side. The first building was constructed on a raft foundation, whereas the second building was on isolated footings, Fig. 6. The two buildings consist of a ground floor and five typical floors and were constructed as reinforced concrete skeletons and in filled brick walls approximately 32 years before.

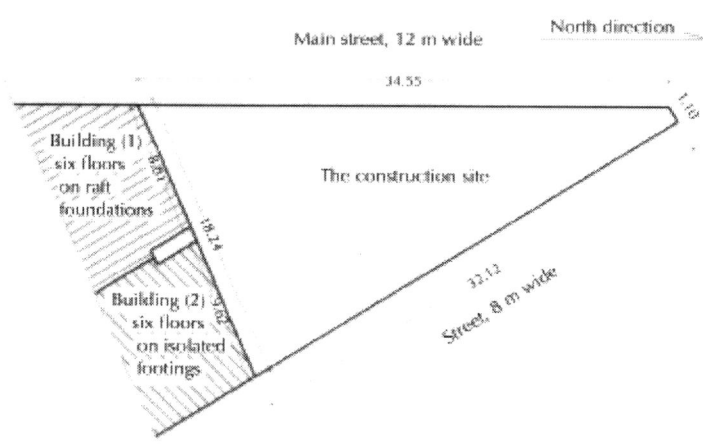

Figure 6: Layout of the construction site in case study #2.

Three boreholes were drilled at the site up to 35.00 m in depth. The recovered soil samples from the boreholes were classified in accordance with ASTM D 2487. Geotechnical investigations showed that the top fill layer comprises crushed sandstone extending to a depth of 2.00 m below the ground surface. The fill overlies soft silty clay layer extending to a depth of 6.00 m underlain by sandy silty up to a depth of 7.00 m. At this depth, poorly graded sand with silt was encountered and extended to a depth of 15.00 m below the ground surface. At this depth, a peat formation of 5.00 m in thickness was explored. The peat is intervened by stiff clay at a depth of 16.00 m and extended to a depth of 18.00 m. At a depth of 20.00 m, a bed of well graded sand trace silt was explored up to the end of the borings. Fig. 7 illustrates a typical borehole log at the site.

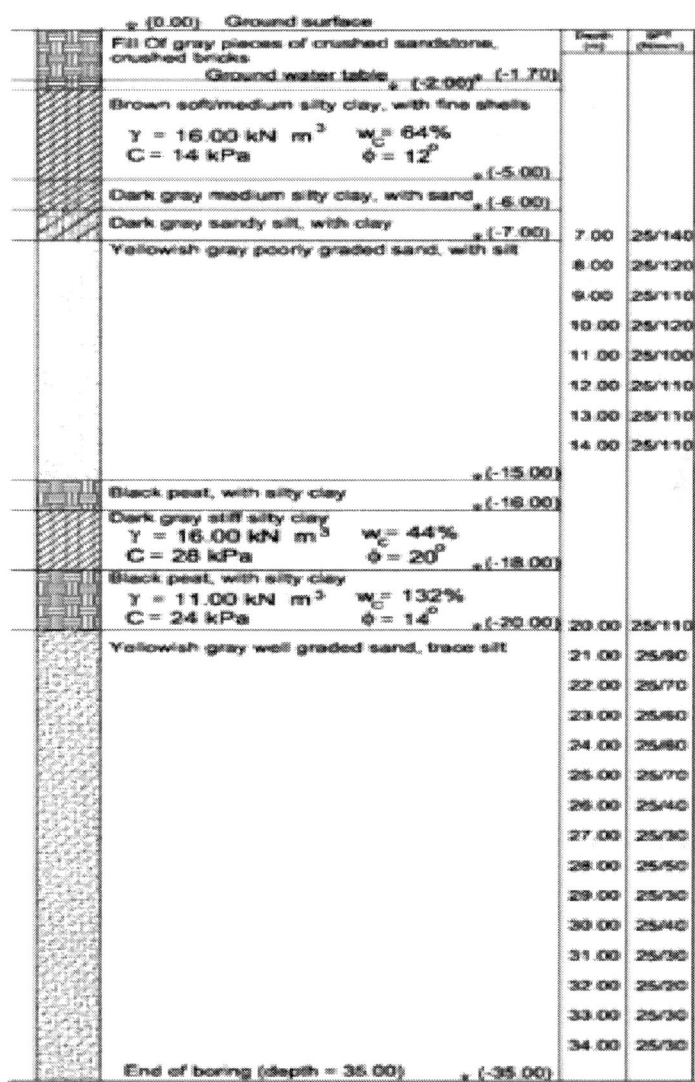

Figure 7: Typical borehole in case study #2.

Side Effects of Pile Installation and the Remedy Procedure

ACIP of 600 mm in diameter and 11.00 m in depth were designed to support the proposed building. The pile working load is 0.65 MN, and the test load is 1.10 MN. Carefully investigations of the two adjacent buildings revealed that they were constructed without implementing any engineering rules, but they appeared stable before work commenced at the new site. The challenge of the engineer was to safeguard these two buildings and to minimize the influence of the pile installation process on the stability of the two buildings. Therefore a row of micro-piles of 150 mm in diameter and 13.00 m in depth below the ground surface was installed adjacent to the existing buildings. Micro-piles of type (B) were used [21]. The spacing between the micro-piles is approximately 250 mm. Cement mortar of 2 (cement):1 (water) was used to form the micro-piles by the grouting technique.

During the construction stage, accurate monitoring of the two buildings using total station of accuracy ±1 mm was recorded. The first building, on a raft foundation, swayed 35 mm towards the construction side at the upper most point of the building, which is 18.00 m above the ground surface. The second building, on isolated footings, swayed 51 mm towards the construction site at a height of 18.00 m. The lateral displacements occurred during the construction of both micro-piles and ACIP. However, 90% of this movement occurred during the construction of the micro-piles.

After installing the micro-piled wall nearby to the existing buildings, the effect of installing ACIP on the stability of the two adjacent buildings was tremendously decreased. The micro-piles wall performed as a barrier to reduce the extent of soil disturbance to the existing buildings, which was observed by the measured rate of tilt of the two buildings. The rate of tilt decreased appreciably after the construction of the micro-piles wall. The additional trivial tilt of the two buildings during ACIP installation may be due to the lateral deformation of the micro-pile wall, which was most likely due to the change in the soil properties in front of the micro-piles wall. The change in the soil properties resulted from soil disturbance caused by ACIP installation. The stability of the micro-piles was based on the measured soil properties according to pre-construction geotechnical investigations. In addition to the existence of micro-piles, additional precautions were taken during installation of ACIP by controlling the rotational and the advance speeds of the auger by using thrust loading on the auger. During drilling process, the developed cyclic shear stresses caused disturbance of soil around the pile hole. As the advance speed of the auger increases, the anticipated soil disturbance decreases. Thrust load can be used to increase the advance speed of the auger.

In another site nearby case study (2), with the same soil conditions, the pile contractor started to install ACIP without any precautions to safeguard the adjacent buildings that were founded on isolated footings about 20 years ago. As a result, the adjacent buildings tilted towards the construction site by 48 mm. The installation process of ACIP was stopped, and a strip area of soil in the construction site and adjacent to the existing buildings was grouted by a cement grout. The depth of the grout extended to 2.00 m below the designed tip level of the ACIP. The grouted soil acted as a barrier to prevent the extent of soil disturbance below the existing buildings. After the grouting process, the tilt of the buildings due to the installation of ACIP decreased appreciably.

Learned Lessons

It can be concluded from this case study that the unfavorable effects of installing ACIP in saturated loose and medium sandy soil can cause tilt of the nearby structures. Thus, it is an important task for the geotechnical engineer to safeguard the nearby buildings during the installation of ACIP. A row of micro-piles and/or grouting of the soil adjacent to the existing buildings using a cement grout were successively used to reduce the side effects of ACIP.

CASE STUDY NO. 3

Site Description and Subsoil Formations

At a construction site located at Alexandria, Egypt, the site is bounded at the northern side by a side road that is 6.00 m in width, Fig. 8. Along this road, there are residential buildings with an expansion joint at 12.00 m from the side road. The buildings are built as a reinforced concrete skeleton. These buildings were constructed 10 years ago and founded on ACIP with a length of 20.00 m below the ground surface. The existing buildings consist of a basement, ground floor, and 11 typical floors and fully occupied.

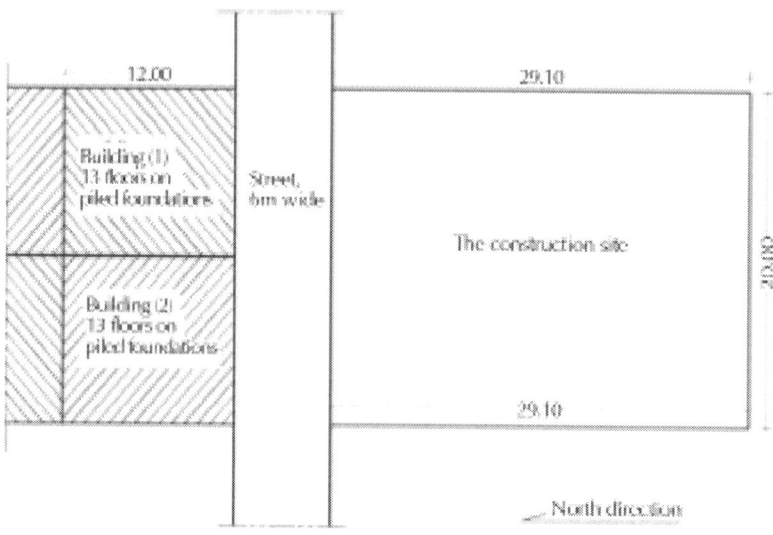

Figure 8: Layout of the construction site in case study #3.

Prior to construction, four boreholes were conducted up to 30.00 m in depth at the site. The recovered soil samples from the boreholes were classified in accordance with ASTM D 2487. The subsoil consists of four successive layers. The top fill layer comprises loamy sand with little crushed sandstone and bricks and extends to depth 8.00 m below the ground surface. The second layer is soft silty clay with fine crushed shells extending down to a depth of 11.50 m below the ground surface. The third layer comprises sandy silt with clay and extends to a depth of 14.00 m below the ground surface. A bed of well graded sand with silt was explored up to the end of borings. Fig. 9 illustrates a typical borehole log at the site. ACIP of 500 mm in diameter and 22.00 m in depth were designed to support the proposed building. The pile working load is 0.90 MN, and the test load is 1.50 MN.

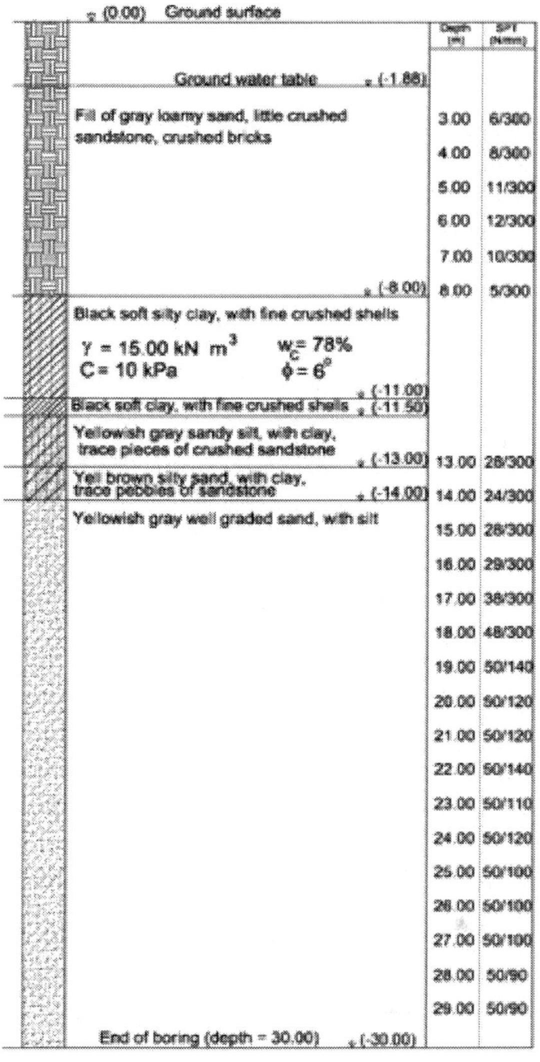

Figure 9: Typical borehole in case study #3.

Effect of Pile Installation on the Adjacent Buildings

The adjacent buildings were stable before work commenced at the construction site. During the construction stage of ACIP, it was noticed that the asphalt level of the side road randomly subsided with a maximum value of 250 mm. Moreover, the existing buildings swayed towards the construction site by about 32 mm, and the expansion joint of the existing building became wider than the designed space. Although the measured sway was within the allowable limits, cracks were observed in slab on grad inside the basement of the existing building just below the expansion joint. It is clear that the existing building that founded on ACIP, similar to the piles installed in the construction site, suffered tilt due to the installation of ACIP. The depth of the piles of the existing building is 20.00 m, whereas the depth of piles installed in the construction site is 22.00 m.

The installation of ACIP for the new building with longer depth than the pile length of the existing building caused cyclic shear stresses in the surrounding soil. The cyclic shear produced high-pore water pressure accompanied by a decrease in the shear strength of the soil. The dissipation of the excess pore water pressure and the lateral movement of soil towards zone of low-shear strength caused soil subsidence around the piles. This subsidence extends away from the center line of the hole by a conical shape. According to Abdrabbo and Gaaver [3], this conical shape has a radius equal to ten times the pile diameter at the pile tip with inclined surface of 4 (vertical): 1 (horizontal). The vertical movements of geomaterial in the construction site impose additional drag down loads on the existing piles. The imposed drag down load depends upon the soil movement adjacent to an existing pile. As the distance from the construction site increases and the soil movement decreases, the imposed drag down load on a pile decreases. As a result, the tilt of the existing building towards the construction site occurred. Also, due to the decrease in shear strength of soil at the construction site, lateral movement of soil towards the construction site occurred. In addition, the existing building imposed lateral pressures on the freshly constructed piles nearby. These lateral pressures cause further lateral movement of soil to the construction site. The lateral movement of the soil may have contributed to the building tilt.

Learned Lessons

It can be concluded from this case study that the unfavorable effects of installing ACIP in saturated loose and medium sandy soil can cause tilt of the nearby structures, even when they are on piled foundations. Thus, the geotechnical engineer is required to safeguard the nearby buildings during the installation of ACIP. More studies are still required to implement a new technology and advance the machinery involved in drilling ACIP in order to decrease the soil disturbance during the installation process.

CONCLUSIONS

Three case studies were presented through this paper in an effort to illustrate learned lessons in auger cast-in-place piles (ACIP). Misuse of ACIP equipment resulting defective piles was discussed through the first case study. The second and the third case studies showed the adverse effects of installing ACIP on the nearby buildings. The study revealed the following

conclusions:

1. Employ a clever pile crew during the installation of ACIP is necessary to observe, interpret, and take corrective actions for unusual situations.

2. The authorities worldwide should oblige pile contractors to employ only experienced and qualified workers in charge of geotechnical engineering works.

3. Tender documents of a project containing ACIP should include precise clauses related to rotational speed of the auger, advance speed of the auger, auger pitch, imposed torque, grout pressure, grout ratio, and thrust load on the auger. The auger should be of height equal to the full depth of the proposed piles, no extension to the auger should be used.

4. The operator of the pile rig should report daily to the engineer the geotechnical and the technological aspects during pile installation in addition to all reordered readings by monitoring equipment. Close and precise supervision of pile construction by an expert in geotechnical engineering is necessary.

5. Employing two pile rigs and disregarding of field observations during pile installation caused instability of the constructed building in the first case study. A remedy procedure including two rows of micro-piles was designed and constructed.

6. Unfavorable side effects of installing ACIP in saturated loose and medium sand can cause tilt of nearby structures; even they are on either shallow or deep foundations.

7. A row of micro-piles and/or soil grouting adjacent to the existing buildings were successfully used to reduce the adverse effects of ACIP.

8. Implementation of different codes on the results of pile loading tests produced different pile working loads. Therefore, tender documents should specify the code upon which conducting tests and interpreting of test results. At the meantime the geotechnical engineer should implement his experience and judgment during application of the specified code.

REFERENCES

1. F.M. Abdrabbo. Effects of imperfections in the construction procedure of auger cast in place piles on pile response, in: Tenth Asia–Pacific Conference, EASEC, Thailand, 2006.

2. F.M. Abdrabbo, H.M. Abouseeda, Effect of construction procedures on the performance of bored piles, in: Proceedings of the International Deep Foundations Congress, 2002, pp. 1483–1454.

3. F.M. Abdrabbo, K.E. Gaaver, Pore water pressure arising during pile drilling in sand, Journal of Civil Engineering and Architecture, USA, 5 (4) (2011), pp. 331–340

4. F.M. Abdrabbo, M.A. Mahmoud, Uncertainties in the performance of flight auger piles, Alexandria Engineering Journal, 32 (3) (1993), pp. C147–C153

5. I. Alpan, The empirical evaluation of the coefficient Ko and Ko, OCR, Soils and foundations, Tokyo, vol. 7, No. 1, 1967, pp. 31–40.

6. L. Bjerrum, I.J. Johannessen, O. Eide, Reduction of negative skin friction on steel piles to rock, in: Proceedings of 7th International Conference soil Mechanics and Foundation Engineering, Mexico, vol. 2, 1969, pp. 27–33.

7. Y.K. Chow, J.T. Chin, S.L. Lee, Negative skin friction on pile groups, International Journal of Numerical Analytical Methods in Geomechanics, 14 (2) (1990), pp. 75–91

8. F. DE Cock, C. Legrand, Design of Axially Loaded Piles-European Practice, Balkema, Rotterdam, Netherlands (1997)

9. DFI, Augered Cast-in-Place Manual, Deep Foundation Institute, Sparata, NJ, 1994, pp. 29.

10. EC, Egyptian Code of Soil Mechanics and Foundations – Deep Foundations Part 4, 2001.

11. P.R. Glass, A.J. Powderham, Application of the observational method at Limehouse Link, Geotechnique (44) (1994), pp. 665–679

12. R.D. Holtz, W.D. Kovacs, An Introduction to Geotechnical Engineering, Prentice-Hall, Englewood Cliffs, NJ (1981)

13. C.J. Lee, C.W.W. Ng, Development of down drag on piles and pile groups in consolidating soil, Journal of Geotechnical and Geoenvironmental Engineering, ASCE, 130 (9) (2004), pp. 905–914

14. C.Y. Lee, H.G. Poulos, Tests on model instrumented grouted piles in offshore calcareous soil, Journal of Geotechnical Engineering, ASCE, 117 (11) (1992), pp. 1738–1753

15. A. Mandolini, M. Ramondini, G. Russo, C. Viggiani, Full scale loading tests on instrumented CFA piles, Proceedings of Deep Foundations (2002), pp. 1088–1097

16. W.J. Neely, Bearing capacity of auger-cast piles in sand, Journal of Geotechnical Engineering, ASCE, 117 (2) (1991), pp. 331–345

17. M.W. O'Neill, K.M. Hassen, Drilled shafts effects of construction on performance and design criteria, in: Proceedings of International Conference on Design and Construction of Deep Found, vol. 1, 1994, pp. 137–187.

18. H.G. Poulos, Pile behavior–consequences of geological and construction imperfections, Journal of Geotechnical and Geoenvironmental Engineering, ASCE, 131 (5) (2005), pp. 538–563

19. H.G. Poulos, Pile defects–influence on foundation performance, in: 4 International Conference Deep Foundations Practice, Singapore, 1999, pp. 57–96.

20. G. Poulos, H. Davis, Pile foundation analysis and design, John Wiley & sons, Inc., New York, NY (1980)

21. V.R. Schaefer, Ground Improvement, Ground Reinforcement, and Ground Treatment – Development, 69Geotechnical Special Publication (1997)

22. C. Viggiani, Further experiences with auger piles in Naples area, in: Proceedings 2nd International Symposium on Deep Foundations on Bored and Auger Piles, Balkema, Rotterdam, 1993, pp. 445–455.

23. C. Vipulanandan, O. Guvene, M. McClelland, Monitoring the installation and curing of a large diameter ACIP in very dense sand, in: Proceedings of GeoDenver, ASCE, 2007.

24. T.H. Wu, 2008 Peck lecture: the observational method: case history and models, Journal of Geotechnical and Geoenvironmental Engineering, ASCE, 137 (10) (2011), pp. 862–873

25. P.P. Xanthakos, Slurry Walls as Structural System, Mc-graw Hill Inc (1994)

Performance of BFRP Retrofitted RCC Piles Subjected to Axial Loads

Anandakumar Ramaswamy,[1] Selvamony Chachithanantham,[2] and Seeni Arumugam[3]

[1]Anna University, Chennai, Tamil Nadu, India

[2]Department of Civil Engineering, Sun College of Engineering &
Technology, Tamil Nadu, India

[3]Department of Civil Engineering, S.V.C. College of Engineering &
Technology, Tamil Nadu, India

ABSTRACT

This paper deals with the behaviour of basalt fibre reinforced polymer
(BFRP) composites retrofitted RCC piles subjected to axial compression

loads. Currently the awareness of using FRP increases rapidly in engineering fields and also among public. Retrofitting becomes vital for aged and damaged concrete structures, piles, and so forth, to improve its load carrying capacity and to extend the service life. The load carrying capacity of piles retrofitted with basalt unidirectional fabric was studied experimentally. 15 nos. of RCC end bearing pile elements were cast with same reinforcement for axial compression experiment. Three piles were used as conventional elements, another 3 piles were used as double BFRP wrapped pile elements, and remaining 9 piles were used as retrofitted piles with BFRP double wrapping after preloaded to 30%, 60%, and 90% of ultimate load of conventional element. The effects of retrofitting of RCC pile elements were observed and a mathematical prediction was developed for calculation of retrofitting strength. The stress vs. strain relationship curve, load vs. deformation curve, preloaded elements strength losses are tabulated and plotted. Besides, crack patterns of conventional elements and tearing BFRP wrapped elements were also observed. The BFRP wrapped elements and retrofitted elements withstand more axial compressive load than the conventional elements.

INTRODUCTION

Generally the RCC structures are degrading due to corrosion, multiple environmental effects, poor quality of construction, deterioration and damages, age, fatigue, increment of live loads, and so forth. Particularly RCC pile foundation shafts and caps are frequently affected and damaged by any one of the above causes. These structures can be strengthened or retrofitted by using the best suited method from jacketing, overlaying, stitching, grouting, sealing, coating, NSM systems, FRP wrapping, rebaring, blanketing, and so forth. FRP wrapping system is well suited for rapid retrofitting works without the application of heavy tools and skilled labours. It won good characteristics such as high chemical and heat resistance, low fatigue loss, high flexibility, high tensile strength, light weight, high impermeability and corrosion free because it is non metallic. Hence, the FRP wrapping method was chosen for this experiment by using basalt unidirectional fabric. The aim of the research study is to assess the behaviour and performance of retrofitted pile elements subjected to axial compression load, impact load, lateral load, and skin friction. This paper mainly focuses on the behaviour of BFRP wrapped piles under the axial compression load. For this study, it is considered the slender piles only, because the piles are slender members, that is, the ratio between effective length and least lateral dimension is greater than 12. As, the load carrying capacity of slender piles are less than that of short piles and various properties like flexural buckling, torsional buckling, strain, and so forth are very much less, when compared to the short piles. The slender piles are considered for this research study. In relation to that, some research articles were reviewed for this experiment and expressed.

Purushotham Reddy et al. experimented RCC piles that is strengthened with GFRP composites under various loading conditions and compared the experimental results with analytical results obtained using ANSYS. They observed that the load carrying capacity of the elements retrofitted with GFRP is found to be greater than that of the control elements for ultimate axial compression and lateral loads at various levels [1].

Olivova and Bilcik evaluated the performance of near surface mounted (NSM) laminates with CFRP wrapped columns using both experimental and analytical methods. The adoption of the NSM technique by using precut grooves in concrete cover resulted in significant increase of load carrying capacity of columns subjected to bending. Some of the experimental researches were carried for specimens under cyclic loadings and seismic loading. However, most of the researches show the performance of specimens under static axial loads and static lateral loads [2].

Abbasnia and Holakoo presented the stress-strain behaviour of CFRP confined concrete under monotonic and cyclic compression. They found that the existing models such as Shao model and Lam and Teng model have less accuracy in predicting the stress, strain, and unloading path [3].

The effects of cyclic loading and the behaviour of the confined concrete in hybrid DSTCs (hybrid FRP-concrete-steel double skin tubular columns) under cyclic axial compression were studied by Yu et al. It was found that the repeated unloading and reloading cycles result in cumulative effects on the permanent stress and strain deterioration of the confined concrete [4].

Desprez and Mazars presented a new simplified modeling strategy for reproducing the nonlinear cyclic behaviour of FRP retrofitted RC columns. This can serve as a numerical tool for quick comparative studies on various confinement situations and structure vulnerability before and after FRP retrofitting [5].

Ozbakkaloglu and Louk Fanggi conducted experimental study on the behaviour of FRP-HSC steel composite columns subjected to monotonic and cyclic axial compression. They made experiment investigation on the double-skin tubular columns and elaborated that the compression behaviour is governed by the concrete strength, steel tube diameter, loading pattern, thickness, and end conditions [6]. Ferracuti and Savoia introduced cyclic constitutive laws for confined and unconfined concrete under compression and concrete under tension and also the steel reinforcing bars. The effects of the axial loads on the damping factor were studied [7].

The strengthening configuration of columns subjected to combined axial load simulating gravity load and reverse cyclic lateral load simulating earth quake load were applied by Sadone et al. on CFRP confined columns with bonded longitudinal CFRP plates [8].

Mirmiran et al. conducted their paper that field experiments have shown concrete filled tubes to be a feasible alternative for bridge substructures. The driving stresses in filled tubes were comparable to those for prestressed concrete piles. Empty tubes may buckle or rupture under driving impact, unless driven at shallow depths and in soft soil or with a steel mandrel. The parametric study using the wave equation further confirmed that there is no difference in the drivability of filled FRP tubes and prestressed concrete piles of the same cross sectional area and concrete strength [9].

Fam et al. detailed that the construction and driving of piles, comparisons between the behaviour of the composite (circular cross section) and prestressed concrete piles (square cross section) under axial and lateral loading, the absorbed failure modes, and the details of connection between the piles and the reinforced concrete cap. The use of concrete filled GFRP tubes as piling for bridge piers is practical and feasible. Both the composite and prestressed concrete piles performed similarly under the axial load test and also the failure due to axial load was significantly higher than the design pile load. In addition, the lateral load field test on both the composite and prestressed piles showed similar behaviour to that attained from the laboratory flexural test and analysis [10].

Parvin and Brighton showed the collective usage of FRP in strengthening columns under several loading exposures. Although, there are numerous amount of FRP strengthening techniques, the BFRP composites wrapping proves to be one of the best retrofitting method [11]. The current study deals with the determination of behaviour and performance of BFRP composites wrapped piles subjected to axial load.

OBJECTIVES

The main objectives of this paper are

- to retrofit the pile grade loss elements by using basalt fibre,
- to find out the compressive strength, split tensile strength and flexural strength of BFRP retrofitted specimens in single and double wrapping,
- to study the performance of BFRP double wrapped single end bearing RCC piles under axial loading,
- to determine the stress versus strain, load versus deformation behaviours, cracks patterns and to compare with conventional pile elements,
- to determine the strength of retrofitted elements,
- to compare the load carrying capacity of conventional pile specimen and retrofitted pile specimen based on IS 2911-Part 1/ Sec 3-1984.

MATERIALS AND METHODS

Experiment Test Materials

For this experiment, the following ingredient such as 43 grade OPC cement, fine aggregate, coarse aggregate, water, steel, basalt unidirectional fabric, and epoxy were used for specimens like cube, cylinder and prism, and end bearing RCC pile element casting and retrofitting purposes.

The ingredients were tested as per Indian standard codes for finding out the properties. Cement was tested as per IS 8112-1989, aggregates were tested as per IS 383-1970, and steel testing was made as per IS 1786-1985. After testing the ingredient properties were obtained and tabulated under Tables 1, 2, and 3.

Table 1: Properties of cement

Test conducted	Result
Specific gravity	3.15
Fineness (specific surface) in m^2/kg	227.8
Initial setting time in minute	45
Final setting time in minute	585
Standard consistency in %	36

Table 2: Properties of fine aggregate

Test conducted	Result
Specific gravity	2.8
Fineness modulus	3.1
Water absorption in %	0.5
Surface texture	Smooth
Particle shape	Angular

Table 3: Properties of coarse aggregate

Test conducted	Result
Specific gravity	2.8
Fineness modulus	7.5
Water absorption in %	0.5
Surface texture	Smooth
Particle shape	Angular
Impact value in %	15.2
Crushing value in %	18.6

Table 4: Basalt unidirectional fabric specifications

Basalt fabric code	Structure weaving	Weight (g/m²)	Thickness (mm)	Width (mm)	Weft and warp density Ends/10 mm	
					Warp	Weft
BWUD-450	Unidirection (UD)	450	0.36	600	3.5	0

In this research the RCC piles are wrapped with basalt fibre by using the matrix of epoxy resin (Araldite LY 556) and hardener (Aradur 5021) in the ratio of 1 : 0.15. The epoxy resin and hardeners properties are got from Araldite company data sheet; it is noted in Table 5.

Table 5: Key data for Epoxy-Araldite LY 556 (resin)

Description	Araldite LY 556 (resin)	Aradur 5021 (hardener paste)
Aspect (visual)	Clear, pale yellow liquid	White viscous paste
Viscosity at 25°C (ISO 9371 B)	10000–12000 [mPa s]	70000–90000 [mPa s]
Density at 25°C (ISO 1675)	1.15–1.2 [g/cm³]	1 [g/cm³]

Based on the ingredient test results it has been decided to design the grade concrete mix as IS 10262-2000 for specimens and elements casting and testing for this experiment.

Experimental Test

Initially the concrete characteristic strengths were observed from conventional concrete specimens such as cubes, cylinders, and prisms as recommended in IS 516-1959. The experimental results and characteristics are noted in Table 6.

Table 6: Conventional specimen's characteristics

Conventional concrete specimen	Geometry in mm	Ultimate compressive strength		Ultimate split tensile strength		Ultimate flexural strength	
		Load in kN	Stress in N/mm²	Load in kN	Stress in N/mm²	Load in kN	Stress in N/mm²
Cube		803	35.7	...	...	...	...
Cylinder	150 ø & 300 length	...	...	280	3.97	...	...
Prism length ...		...	...	...	10.80	5.4	

Nine specimens were wrapped with basalt unidirectional fabric by using epoxy composites in single layer and another 9 specimens were wrapped in double layer. During wrapping the orientation of fibre warp playing an important role in load carrying capacity. For this experiment the fibre warps are made along the circumference (hoop tension direction) for compressive strength, it shows in Figure 1. It proves to be effective when the overlapping of fibre wrapping is more than 25% of the specimen circumference. Figure 2 shows the orientation of fibre for flexural and tension strength warp is made along the vertical direction.

Figure 1: Orientation of fibre for compression warp in horizontal direction.

Figure 2: Orientation of fibre for flexural and tension warp in vertical direction.

For retrofitting experiment purposes, the 3 conventional specimens were preloaded to 30% of mean ultimate compressive strength of conventional specimens. Similarly, another 6 specimens were preloaded to 60% and 90% of ultimate load on each 3 specimens. These preloaded specimens are retrofitted with BFRP single wrapping for determining its retrofitting performance. Similarly moreover, another 9 nos. of cubes were preloaded and retrofitted with BFRP double wrapped for the same.

Double wrapped specimens showed more strength than conventional specimens and single wrapped specimens as shown in Figure 3. This is due to the high hoop tensile strength of double wrapped specimens than the single wrapped specimens and conventional specimens. Based on this results further experiments were made on double BFRP wrapping systems as the double wrapping specimens possess high compressive strength, split tensile strength and flexural strength than conventional and single wrapping specimens.

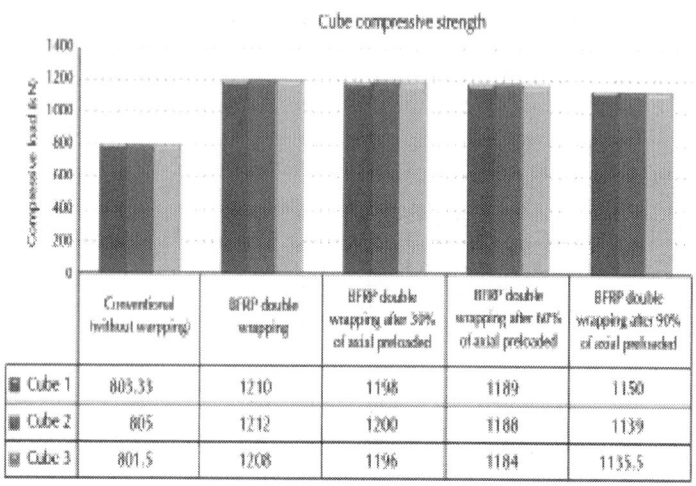

	Conventional (without wrapping)	BFRP double wrapping	BFRP double wrapping after 30% of axial preloaded	BFRP double wrapping after 60% of axial preloaded	BFRP double wrapping after 90% of axial preloaded
■ Cube 1	803.33	1210	1198	1189	1150
■ Cube 2	805	1212	1200	1188	1139
■ Cube 3	801.5	1208	1196	1184	1135.5

Figure 3: Conventional and BFRP retrofitted cubes performance compression load.

From the experiment results the mechanical properties were observed, analysed and plotted. A mathematical prediction equation was arrived for the compressive strength determination of retrofitted cube specimens.

Determination of Constants C_c, k_1, and R_c For Compression Stress Cube

The following data were observed form this experiment.

Grade of concrete mix is M_{30}, characteristics compressive stress is

$$f_{ck} = 30N / mm^2$$

(1)

Experimental concrete cube characteristics compressive stress f_{cke} is

$$f_{cke} = k_1 f_{ck} \tag{2}$$

where k_1 is a constant and its value is arrived as 1.19 for all design calculation.

Characteristics compressive stress f_{ckw} of BFRP double wrapped concrete cube is

$$f_{ckw} = f_{cke} C_c, \tag{3}$$

Where C_c is a constant and whose value is arrived as 1.5 for all design calculation.

Retrofitted Cubes. Retrofitted concrete cube strength

$$f_{ckr} = f_{ckw} R_c{}'$$

$$f_{ckr} = k1 f_{ck} C_c R_c. \tag{4}$$

BFRP double wrapped cube specimens preloaded by strength 30% of the compressive strength of conventional cube is calculated from constants as follow:

$$f_{ckr} = f_{ckw} R_c \; or \; f_{ckr} = k1 f_{ck} C_c R_c \tag{5}$$

R_c value is taken as 0.994 from graph shown in Figure 4.

Ultimate stress of retrofitted specimens

$1.19 \times 30 \times 1.5 \times 0.994$

$$= 53.22 N / mm^2 \tag{6}$$

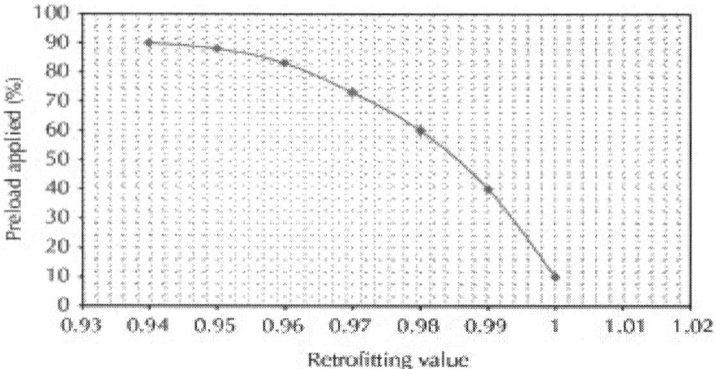

Figure 4: Retrofitting Constant Value for compressive strength.

BFRP double wrapped cube specimens preloaded by strength 60% of the compressive strength of conventional cube is calculated from constants as follows:

$$f_{ckr} = f_{ckw}R \tag{7}$$

or

$$f_{ckr} = k1 f_{ck} C_c R_c \tag{8}$$

R_c value is taken as 0.98 from graph shown in Figure 4. Consider Ultimate stress of retrofitted specimens

$1.19 \times 30 \times 1.5 \times 0.98$

$$= 52.47 N / mm^2 \tag{9}$$

BFRP double wrapped cube specimens preloaded by strength 90% of the compressive strength of conventional cube is calculated from constants as follows:

$$f_{ckr} = f_{ckw}R_c \tag{10}$$

or

$$f_{ckr} = k1 f_{ck} C_c R_c \qquad (11)$$

R_c value is taken as 0.94 from graph shown in Figure 4. Consider Ultimate stress of retrofitted specimens

$1.19 \times 30 \times 1.5 \times 0.9$

$$= 50.33 N / mm^2 \qquad (12)$$

Element Casting and Testing

The aim of this experiment is to test the retrofitting of damaged RCC end bearing piles using BFRP composites wrapping method. For this experiment the pile elements ultimate load carrying capacity of the pile elements 379 kN was designed as per IS 456-2000 and IS 2911-Part 1/Sec 3-1984. 15 pile elements were cast with uniform geometry, mechanical properties and reinforcements as shown in Table 9 and Figure 5 for testing purposes under different loading conditions as per Table 10 schedule.

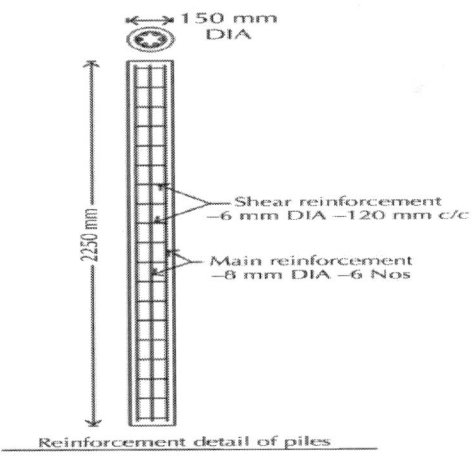

Figure 5: Reinforcement details of Pile elements.

The conventional pile elements (designated as AC 1, AC 2, and AC 3) were prepared for testing with following procedure. The pile elements were coated with white cement after cleaning of pile shaft; white cement wash is used for grids line marking and cracks pattern studies. Then the pile elements were fixed vertically in the 2000 kN capacity loading frame. Besides, mechanical strain gauges were fixed at three places such as 1/3rd, middle, and 2/3rd spans interval for observing the deformation of elements during compressive loading. Two dial gauges were fixed at middle of elements on right side and back side for measuring the deflection of elements. Initial readings were observed during free loading condition. Then compressive load is applied at 50 kN series interval up to ultimate carrying capacity of pile elements. For every 50 kN loading intervals strain and dial gauge readings were observed and cracks formations places, patterns, width, and so forth were studied. The conventional pile elements fails at the middle due to compression as shown in Figure 6.

Figure 6: Conventional elements testing.

Similarly, the BFRP composites double wrapped elements were tested as per Table 10 schedule. In this, the orientation of fibre warp is made along the circumference; it is applicable for withstanding the hoop stress as the fibres won good tensile properties.

For retrofitting procedure, first the conventional pile elements were preloaded to 30% of ultimate compression load on three elements then double wrapped using BFRP composites. Similarly another 6 elements were preloaded to 60% and 90% on each 3 numbers before double wrapping with BFRP composites. Then elements were tested under the axial compressive loads.

The conventional pile elements were designed for an ultimate load carrying capacity of 379 kN. But, in this experiment the piles elements carried an ultimate mean load of 400 kN. The BFRP double wrapped elements withstand a mean ultimate load of 932 kN. The retrofitted elements preloaded to 30%, 60%, and 90% withstand mean ultimate load of 922 kN, 912 kN, and 890 kN, respectively.

The deformation readings were observed from the mechanical strain gauge during the experiment for every 50 kN. From that reading the stress and strain values are found and plotted as shown in Figure 9.

RESULTS AND DISCUSSIONS

In this experiment, the performance and behaviour of BFRP wrapped and retrofitted elements were observed under the axial compressive load. The BFRP wrapped elements performed well in compression loading conditions by enhancing the load carrying capacity, resistance to deformation, and deflection than the conventional elements. Comparisons are as follow.

1. From the initial test results shown in Tables 7 and 8, it is clear that the BFRP double wrapped cube specimens possess 129% and 150% more axial compressive strength than single wrapping specimens and conventional specimens, respectively. Double BFRP wrapped cylinders possess 130% and 206% more split tensile strength than single wrapping specimens and conventional specimens, respectively. Similarly double BFRP wrapped prisms possess 109% and 223% more flexural strength than single wrapping specimens and conventional specimens, respectively. This is due to the unidirectional fabric wrapping which induces the strength of the specimens. Fibre warp made in the circumference withstands more hoop stress; thus, the BFRP wrapped specimens shows more strength.

2. As shown in Figure 3 retrofitted specimens possess 149%, 147%, and 142% more strength than conventional specimens. Generally fibres have good flexural and tensile strength than steel. Relative to that, the elements double wrapped with BFRP composites for retrofitting gives more compressive strength, high tensile strength and stiffness, leading to measure in load carrying capacity.

3. In Figure 8, the load carrying capacity of pile elements were compared with conventional pile elements. The BFRP wrapped elements attained 2.33 times more than conventional elements. Similarly, 30%, 60% and 90% preloaded then retrofitted pile elements showed enhancement in the compressive strength for about 2.3%, 2.27%, and 2.25% times than conventional elements, respectively. This enhancement of compressive strength is caused by the wrapping of BFRP along the hoop direction which won good mechanical properties.

4. The stress versus strain relation between BFRP wrapped and conventional pile elements was exposed in Figure 9. BFRP double wrapped elements remained stable than conventional elements during the axial compressive loading. In addition to that the volumetric strain was found to be lower than conventional elements. Thus, BFRP wrapping supports the elements to withstand and reduce its deformation under compressive loading.

5. Cracks were initially formed at mid span of the conventional elements. First fine crack was formed at mid span on 225 kN compressive loading. Thin cracks were formed at mid span and adjacent of the mid span on 250 kN. Medium cracks were formed at mid span and adjacent of the mid span on 300 kN. Medium cracks get widened on 350 kN. Width of the cracks increases abruptly on ultimate loading that is 400 kN as shown in Figure 6. BFRP composites wrapped piles get buckled and torn at ultimate loading as shown in Figure 7.

6. Strength of Retrofitted specimens were found using mathematical prediction and compared with experimental results as shown in Table 8. The mathematical prediction results show negligible deviations with experimental results.

7. The load carrying capacity of conventional pile specimens and retrofitted pile specimens were compared as per IS 2911-Part 1/ Sec 3-1984.

Table 7: BFRP wrapped concrete specimens characteristic stress

Specimens	BFRP composites wrapping layer	Characteristics stress in N/mm²
Cube	Single	41.4
	Double	53.77
Cylinder	Single	6.25
	Double	8.18
Prisms	Single	11.05
	Double	12.08

Table 8: Concrete specimens compressive stress comparison between experimental results and mathematical prediction

Description of concrete cube specimens	Compressive stress in N/mm²	
	Experimental results	Mathematical prediction
Conventional	35.7	35.7
BFRP double wrapped	53.77	53.77
30% preloaded of ultimate compressive load then BFRP double wrapped	53.50	53.22
60% preloaded of ultimate compressive load then BFRP double wrapped	52.75	52.47
90% preloaded of ultimate compressive load then BFRP double wrapped	50.75	50.33

Table 9: Properties of RCC end bearing piles

Diameter of pile	150 mm
Height of pile	2250 mm
Slenderness ratio	15
Main reinforcement	8 mm Ø RTS rod—6 nos. vertically
Shear reinforcement	6 mm Ø RTS rod—120 mm c/c spacing
% of reinforcement	1.706
Grade of concrete	M 30

BFRP wrapping thickness in two layers	2.04 mm
Basalt fibre orientation	Warp comes along the circumference direction

Table 10: Schedule of RCC pile elements testing program

Sl. No.	Description of element	Designation of element	Nos. element used for testing
1	Conventional element	AC 1, AC 2, and AC3	3
2	BFRP composites double wrapped element	ACW1, ACW2, and ACW3	3
3	30% preloaded of conventional element ultimate load the BFRP double wrapped	AR3-1, AR3-2, and AR3-3	3
4	60% preloaded of conventional element ultimate load the BFRP double wrapped	AR6-1, AR6-2, and AR6-3	3
5	90% preloaded of conventional element ultimate load the BFRP double wrapped	AR9-1, AR9-2, and AR9-3	3

Figure 7: BFRP double wrapped elements testing.

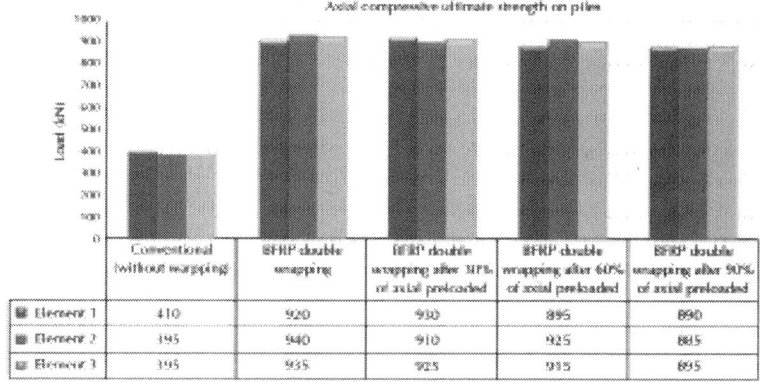

Figure 8: Experimental results of RCC pile elements under the compression.

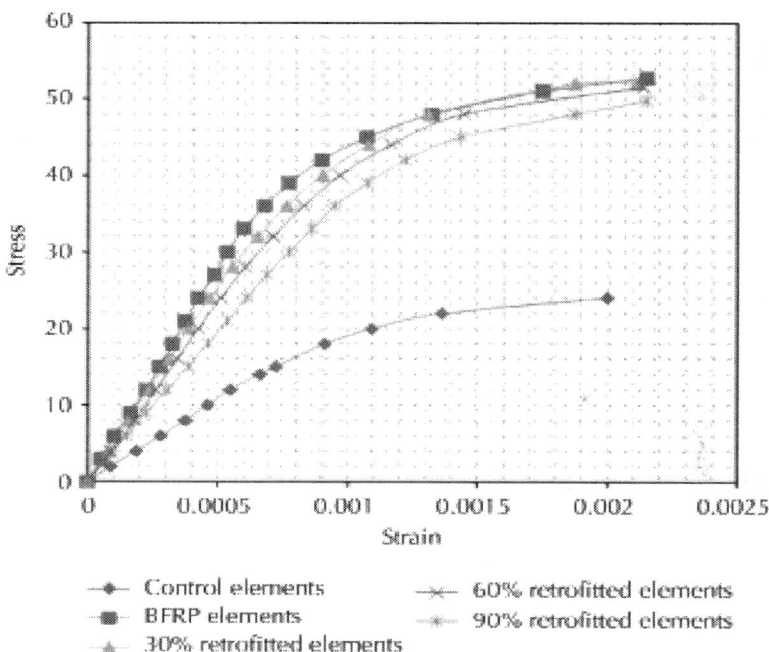

Figure 9: Stress versus Strain Curve for axial compression loaded pile elements.

CONCLUSIONS

From the research results it is concluded that BFRP wrapped pile elements withstand more load carrying capacity, resistance to deformation, and deflection than the conventional elements.

The BFRP double wrapped cube specimens carried more axial compressive strength than single wrapped specimens and conventional specimens. Similarly double wrapped prisms and cylinders were more load carrying capacity than single wrapped and conventional specimens.

The double wrapped element with BFRP composites for retrofitting gives more compressive strength, high tensile strength, and stiffness.

Retrofitted piles after 30%, 60%, and 90% preloaded pile elements showed more enhancement in compressive strength than conventional elements. The enhancement of compressive strength is due to the wrapping of BFRP in the hoop direction.

BFRP double wrapped elements remained more stabled than the conventional elements during the axial loading. The volumetric strain is lower than conventional elements.

The formation of the crack and the crack width on BFRP wrapped pile elements was at very high load ranges when compared to the conventional elements.

Further research can be done to determine the various types of the loading such as lateral load, impact load, and so forth with varying the slenderness ratio. The further works can be extended for finding out the values of durability aspects. This research can be extend for finding out the load carrying capacity of the various types of piles like friction pile, friction, and end bearing pile.

Retrofitting of piles with BFRP wrapping results in tremendous increase of strength parameters and performance of piles. BFRP wrapping endures corrosion and increase the life of the elements. Thus, it can be concluded that BFRP wrapping can be used for retrofitting of members where compression prevails.

REFERENCES

1. B. Purushotham Reddy, P. Alagusundaramoorthy, and R. Sundaravadivelu, "Retrofitting of RC piles using GFRP composites," KSCE Journal of Civil Engineering, vol. 13, no. 1, pp. 39–47, 2009.

2. K. Olivova and J. Bilcik, "Strengthening of concrete columns with CFRP," Slovak Journal of Civil engineering, vol. 17, no. 1, pp. 1–9, 2009.

3. R. Abbasnia and A. Holakoo, "An investigation of stress-strain behavior of FRP-confined concrete under cyclic compressive loading," International Journal of Civil Engineering, vol. 10, no. 3, pp. 201–209, 2012.

4. T. Yu, Y.-B. Cao, B. Zhang, and J. G. Teng, "Hybrid FRP-concrete-steel double—skin tubular columns: cyclic axial compression tests," Research Online, 6th International Conference on FRP Composites in Civil Engineering, pp. 1–8, 2012.

5. C. Desprez and P. J. Mazars, "Stress-strain model for FRP confined concrete columns under cyclic and seismic loading," in Proceedings of the 15th World Conference on Earthquake Engineering (WCEE '12), Lisbon, Portugal, 2012, http://www.iitk.ac.in.

6. T. Ozbakkaloglu and B. A. Louk Fanggi, "FRP-HSC-steel composite columns: behavior under monotonic and cyclic axial compression," Materials and Structures, pp. 1–19, 2013.

7. B. Ferracuti and M. Savoia, "Cyclic behaviour of FRP-wrapped columns under axial and flexural loadings," in Proceedings of the International Conference on Fracture, Turin, Italy, 2005.

8. R. Sadone, M. Quiertant, J. Mercier, and E. Ferrier, "Experimental study on RC columns retrofitted by FRP and subjected to seismic loading," in Proceedings of the 6th International Conference on FRP Composites in Civil Engineering (CICE '12), Rome, Italy, 2012, http://www.iifc-hq.org.

9. A. Mirmiran, Y. Shao, and M. Shahawy, "Analysis and field tests on the performance of composite tubes under pile driving impact," Composite Structures, vol. 55, no. 2, pp. 127–135, 2002.

10. A. Fam, M. Pando, G. Filz, and S. Rizkalla, "Precast piles for route 40 bridge in Virginia using concrete filled FRP tubes," PCI Journal, vol. 48, no. 3, pp. 32–45, 2003.

11. A. Parvin and D. Brighton, "FRP composites strengthening of concrete columns under various loading conditions," Polymers, vol. 6, no. 4, pp. 1040–1056, 2014.

Citations

CHAPTER 1

M Dithinde; J V Retief,Pile design practice in southern Africa Part I: Resistance statistics, Print version ISSN 1021-2019.

CHAPTER 2

J V Retief; and M Dithinde, Pile Design Practice in Southern Africa Part 2: Implicit Reliability of Existing Practice, ISSN 1021-2019.

CHAPTER 3

W. Elsamee, Evaluation of the Ultimate Capacity of Friction Piles, Engineering, Vol. 4 No. 11, 2012, pp. 778-789. doi: 10.4236/eng.2012.411100.

CHAPTER 4

Ding, J. , Cao, Y. , Wang, W. , Zhao, T. and Feng, J. (2014) Experimental Study of Dynamic Characteristics on Composite Foundation with CFG Long Pile and Rammed Cement-Soil Short Pile. Open Journal of Civil Engineering, 4, 1-12. doi: 10.4236/ojce.2014.41001.

CHAPTER 5

ong-ping Guan, Wen Zhao, Shen-gang Li, and Guo-bin Zhang, Key Techniques and Risk Management for the Application of the Pile-Beam-Arch (PBA) Excavation Method: A Case Study of the Zhongjie Subway Station, 2014. doi:10.1155/2014/275362.

CHAPTER 6

Bingxiang Yuan, Rui Chen, Jun Teng, et al., "Effect of Sand Relative Density on Response of a Laterally Loaded Pile and Sand Deformation," Journal of Chemistry, Article ID 891212, in press.

CHAPTER 7

F.M. Abdrabbo and K.E. Gaaver, Simplified Analysis of Laterally Loaded Pile Groups, doi:10.1016/j.aej.2012.05.005.

CHAPTER 8

Y. Mostafa, "Design Considerations for Pile Groups Supporting Marine Structures with Respect to Scour,"Engineering, Vol. 4 No. 12, 2012, pp. 833-842. doi: 10.4236/eng.2012.412106.

CHAPTER 9

Zuzanka Trojanová, Pavel Luká , Zoltán Száraz, and Zden k Drozd

(2014). Mechanical and Acoustic Properties of Magnesium Alloys Based (Nano) Composites Prepared by Powder Metallurgical Routs, Light Metal Alloys Applications, Dr. Waldemar A. Monteiro (Ed.), ISBN: 978-953-51-1588-5, InTech, DOI: 10.5772/57454.

CHAPTER 10

Fathi M. Abdrabbo, Khaled E. Gaaver, Installation Effects of Auger Cast-in-place Piles, doi:10.1016/j.aej.2012.08.001.

CHAPTER 11

Anandakumar Ramaswamy, Selvamony Chachithanantham, and Seeni Arumugam, Performance of BFRP Retrofitted RCC Piles Subjected to Axial Loads, doi.org/10.1155/2014/323909.

Index